THE CLIMATE QUESTION

THE CLIMATE QUESTION

Natural Cycles, Human Impact,

Future Outlook

Eelco J. Rohling

OXFORD
UNIVERSITY PRESS

OXFORD
UNIVERSITY PRESS

Oxford University Press is a department of the University of Oxford. It furthers the University's objective of excellence in research, scholarship, and education by publishing worldwide. Oxford is a registered trade mark of Oxford University Press in the UK and certain other countries.

Published in the United States of America by Oxford University Press
198 Madison Avenue, New York, NY 10016, United States of America.

CIP data is on file at the Library of Congress
ISBN 978-0-19-091087-7

1 3 5 7 9 8 6 4 2

Printed by Sheridan Books, Inc., United States of America

To mum, dad, and all others dear to me for support while studying and writing about climate change, to my early mentors Jan-Willem Zachariasse and Peter Westbroek for directing me along this path, and to my sons Ewout and Yarno with hope that their generation will make our planet great again; we have but one.

CONTENTS

CONTENTS

ACKNOWLEDGMENTS

I am greatly indebted to my family for tolerating my absences and anti-social hours locked behind a laptop or in dreamland inside my own head. In addition, I thank all friends and colleagues who kept pushing me to reach beyond the specialist community, and to then discuss the project with me until technicalities were reduced to the essential minimum. I will not list names—you all know who you are and what you have contributed. I am very grateful to my editor, publisher, and reviewers for their invaluable feedback throughout the process of completing the manuscript. Finally, I thank my parents and brother for their unwavering support to all my choices, no matter how irrational, for more than half a century—long may it continue!

THE CLIMATE QUESTION

[1]

INTRODUCTION

In 2015, the annual mean global atmospheric carbon dioxide (CO_2) level surpassed 400 parts per million (ppm; Figure 1.1), and we know very well that this rise is caused by human activities (Figure 1.2). It was the first time in 3 million years that such a level had been reached. Crossing this level has caused widespread concern among climate scientists, and not least among those called paleoclimatologists, who work on natural climate variability in prehistoric times, before humans.

Over the last few decades, researchers have been repeatedly raising the alarm that emissions of CO_2, along with those of other greenhouse gases, are getting dangerously out of control and that urgent remedial action is needed. With the crossing of the 400 ppm threshold, this sense of urgency reached a climax: at the Conference of Parties 21 meeting in Paris—also known as COP21 or the 2015 Paris Climate Conference—broad international political agreement was reached to limit global warming to a maximum of 2°C, and if at all possible 1.5°C, by the end of this century. If one calculates this through, this implies a commitment for society to operate on zero net carbon emissions well before 2050, along with development and large-scale application of methods for CO_2 removal from the climate system. (Scientists focus on carbon (C) emissions when they discuss emissions because it helps in calculating CO_2 changes produced by the processing of specific volumes/masses of fossil fuel hydrocarbons.) Clearly, the challenge is enormous, especially given that even implementing all the pledges made since COP21 would still allow warming to reach about 3°C by 2100. But, regardless, the agreement was ground breaking. It was a

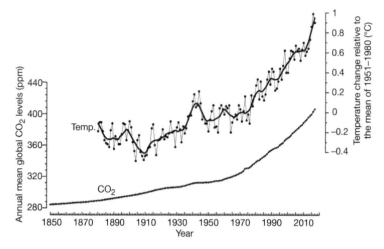

Figure 1.1. Historical changes in key climate factors. Bottom: Annual mean global CO_2 levels.[i] Top: Global average surface temperature.[ii] In the temperature graph, the thick line shows the long-term change by smoothing out short-term variability. The temperature developments over time are extremely similar between six independent assessments, some of which include data from 1850 onward.[iii]

i. *https://www.co2.earth/historical-co2-datasets*

ii. *https://data.giss.nasa.gov/gistemp/graphs/*

iii. *http://berkeleyearth.org/global-temperatures-2017/*

marker of hope, optimism, and international motivation to tackle climate change.

Moreover, there are concerns about the stated COP21 targets. First, the proposed 2°C or 1.5°C limits to avoid "dangerous" climate impacts may sound good, but there is no specific scientific basis for picking these particular numbers.[1] Second, the implied "end of this century" deadline is an arbitrary moment in time. It's not as if the ongoing climate changes will stop at that time, even if we managed to stop all emissions. Instead, change will continue, as well-known slow processes in the climate system adjust to the initial disturbances observed in recent history (Figure 1.1). These slow processes include ocean warming (first the surface, then also

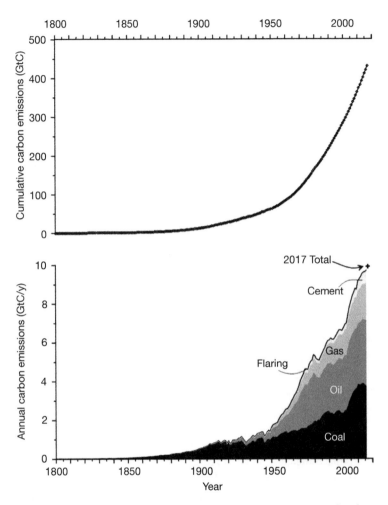

Figure 1.2. Carbon emissions due to human actions. Bottom: Annual carbon emissions from the dominant anthropogenic processes. [iv,v] Top: Cumulative carbon emissions; i.e., the sum total of all preceding years. GtC stands for gigatons of carbon, where a gigaton is one billion (one thousand million) metric tons. The breakdown for 2017 had not yet been released, but the overall total could be estimated from observations. To relate carbon emissions to CO_2, use 1 GtC = 3.67 Gt of CO_2, and note that accumulation of 2.12 GtC in the atmosphere gives a CO_2 rise of 1 ppm. Thus, the top panel shows that we have by now

Figure 1.2. *Continued*

emitted some 430 GtC, or just under 2 times the amount of carbon that we'd calculate from the about 120 ppm rise in atmospheric CO_2 levels (Figure 1). The rest has been mostly absorbed into the oceans (section 4.1).

iv. BP statistical review of world energy, June 2017. https://www.bp.com/content/dam/bp/ en/corporate/pdf/energy-economics/statistical-review-2017/bp-statistical-review-of-world-energy-2017-full-report.pdf

v. Hansen, J., Kharecha, P., Sato, M., Masson-Delmotte, V., Ackerman, F., Beerling, D., Hearty, P.J., Hoegh-Guldberg, O., Hsu, S.-L., Parmesan, C., Rockstrom, J., Rohling, E.J., Sachs, J., Smith, P., Steffen, K., Van Susteren, L., von Schuckmann, K., and Zachos, J.C., Assessing "dangerous climate change": Required reduction of carbon emissions to protect young people, future generations and nature. PLoS ONE, 8, e81648, doi:10.1371/journal.pone.0081648, 2013.

the deep ocean), which takes much more time than warming over land, as well as global continental ice-volume reduction and vegetation changes.

Like a heavy freight train that has slowly gained momentum, these slow processes will continue to unravel over timescales of centuries to millennia, causing total global warming to keep increasing over many centuries. For example, slowly increasing temperatures of the oceans—which cover about 70% of Earth's surface—cause the global average temperature to creep up, while gradual reduction of ice volume slowly but surely makes the planet less reflective to incoming sunlight and thus causes further warming. Vegetation changes also affect the planet's reflectivity, as well as impacting natural carbon exchanges with the atmosphere. And so on. What we see here is that the climate system is a complex beast, which includes the atmosphere, ocean, and biosphere, and all the interactions between them. Over still longer timescales, the climate system even has important interactions with ocean sediments, sedimentary rocks, and volcanic processes.

Because of the slow processes, any climate response measured by the year 2100 will be roughly doubled over subsequent centuries, even if there would be no further emissions.[2] To keep longer-term warming below 2°C, we would therefore need to limit warming to a maximum of 1°C by 2100. Incidentally, we have crossed the 1°C value in 2015 already (relative to the pre-industrial value; Figure 1.1), and ongoing greenhouse gas emissions along with global reductions in particulate air pollution are likely to push us through the 2°C threshold within a few decades.[3] It appears that if we really want to achieve the agreed COP21 targets, carbon removal from the climate system will be essential. This might be done by means of natural processes, artificial (human-made) processes, or both.

This scenario illustrates how much of the climate debate can be summarized in one straightforward question: *What governs natural climate cycles and what can nature do in the future,* versus *what is the human impact on climate and what can we do in the future?*

This book explores the question from a perspective of research on past climates, or paleoclimate research. Paleoclimate research is essential for an understanding of what nature can do. In specific terms, we need paleoclimate research to see the longer-term consequences of CO_2 levels of 400 ppm or more, given that one has to go back as much as 3 million years to find a previous period with such high CO_2 levels. In addition, this research sheds light on the scale of today's issues relative to changes that occurred before humans were around—in other words, it reveals the context of natural cycles that human impacts are superimposed upon. And finally, paleoclimate studies help us understand to what extent nature can or cannot help us with carbon removal, by revealing her various processes to do so and their timescales of operation. Thus, paleoclimate studies provide critical information about the size of the challenges facing us.

I have been working in paleoclimatology for about 30 years, and mine is a story that most of my colleagues will recognize. As soon as we mention that we work on climate change from a perspective of changes before humans entered the scene, we are subjected to a barrage of questions. More often than not, these interrogations seem targeted toward getting affirmation on three closely related issues. These are (1) that humans cannot possibly be changing the global climate; (2) that all that's happening today is just part of a natural cycle; and (3) that even if we did have some influence, nature has gone through larger climate cycles and therefore will somehow "fix" the problem.

To me, these issues seem based on an assumption or hope that things will somehow sort themselves out. But it is hard to get specific answers about the basis for this assumption. There seems to be a lot of confusion in people's minds about what processes are at play, how strong and fast they are, how they work together, and how the human influence may scale relative to those processes.

The first source of confusion seems to be that the world is such a big place that people cannot imagine that humans could even pretend to influence it on a global scale. But this clearly is a mistake. With some reflection, most will recognize that we have a record of affecting several fundamental issues on a global scale. And for some of these, the realization that this was the case has led to successful action to reverse the damage. A good example concerns stratospheric ozone levels, which were being reduced due to release of refrigerants and gases used to pressurize sprays, increasing our exposure to ultraviolet rays and thus the risk of skin cancer. Once this was realized, remedial action was agreed upon on a global scale, and now the situation is improving. Another example is commercial whaling. Once it was realized that soon there would be not a whale left in the global ocean, broad international action was taken, and now whale numbers are slowly recovering. And then there is the case of lead pollution from additives in petrol and paint. Once that was recognized as a problem, alternatives were developed and lead pollution now is strongly reduced. If you think about it, then you will find more good examples, many with proof that collective international action can and will drive improvements. Why would climate be different?

A second cause of confusion seems to be that most people cannot see our human influence within the proper context of natural changes. They know almost intuitively that nature has—before humans—gone through much larger extremes than what we're talking about for modern climate change, from ice ages to the warm and lush climates of the time of the dinosaurs. But taking the next step, developing a feel for the importance of our human influence in the context of these large natural changes, requires some study and some math.

Unfortunately, a lot of the key points to read up on are widely dispersed in the literature, although many good summaries can be found with some effort. Also—importantly—most people feel insufficiently qualified, or simply do not have the ambition, to do the math. A lot of this is because a myth seems to have developed that one needs sophisticated climate models, and to get everything exactly right, before an informed opinion can be reached. That is simply not true as long as the objective is to just get a sound feel for our impact (it is true when targeting specific forecasts). The topic may be complex in detail, but the big lines are quite straightforward. Most of the sums needed for a decent impression are no more complicated than those needed to balance a household budget.

From experience at parties, meetings, flights, train rides, and interviews, I have learned that most people are quite comfortable with simple sums that illustrate the main thrust of the argument, along with the simplest of graphs and diagrams that I would usually draw on paper napkins and tablecloths (and once on a quality textile napkin). I have long considered setting these arguments out in a text that is accessible to all who are up for a little challenge. It would bring a range of aspects together in one coherent flow, with examples, and steer clear of details. This book is the result.

The general, overview-like approach that I take in this book conflicts a bit with my scientific background. Details, caveats, and uncertainties are essential fodder for students and specialists in the field. In consequence, we have grown used to sprinkling them throughout all of our communications. But in general life, they only cloud the message. Most of us sufficiently grasp the concept of powered flight that we can enjoy a plane voyage without understanding the avionics and aerodynamics involved, and we are happy to leave those to the pilots and engineers. In a similar way, everyone can reach a reasonable and well-informed personal opinion about the state of the climate from a straightforward presentation of the big lines.

The big lines refer to the context of real-life observations that directly illustrate the overall case and its broad implications. This is the level I seek here, avoiding the confusion of models and their different philosophies and assumptions, even though those are absolutely essential for technical arguments. As a result, this book leans heavily on observations from geological history, which most people can intuitively relate to. It underrepresents the depth of process understanding that has come from climate modelling because that level of detail is not needed here. This is purely a pragmatic choice—as they say: "horses for courses."

The book aims to lead the reader through the entire journey: from first principles, through key processes and long-term observational context, to implications. There are many texts that cover one or more of these aspects, but not many span the entire journey in a concise manner without too many caveats and details. And still the climate debate rages on—and the old arguments about natural variability keep cropping up.

So here's my attempt to take you through the entire journey in straightforward terms, with minimal complications. I add diagrams that—similar to my arguments—deliberately focus only on the main points, keeping them in line with the diagrams I often draw on napkins to support my arguments. I support the text and diagrams with citations to the more scientific, complete information published elsewhere.

My aim is to empower you, the reader, for making your own informed decisions about whether we need to push for action on greenhouse gas emissions. I hope this book will provide the necessary background information, in simple terms and with examples, to convince you about that.

To me personally, it's clear that we have arrived at a critical time for taking action toward transforming society—to break our carbon addiction. Future generations may still come to see us as polluting dinosaurs, but I hope it will be as the last lot of them, who finally realized what was coming and took action to avoid it. Let's see if you agree with me at the end of the book or not. At least the arguments will be transparently laid out for all to see.

The key issue we'll be addressing is the well-established rise of global temperatures by roughly 1°C since the start of the industrial revolution, more than 150 years ago, with most of the warming concentrated in the last six decades (Figure 1.1). Several independent monitorings of the global temperature change are in close agreement (see[4] and note 3 of Figure 1.1), all showing almost year upon year of record-breaking heat in the last four years.[5] The year 2016 comprehensively outclassed all previous years, with 2017 and 2015 hot on its heels (Figure 1.1). There is no denying that there is a strong warming trend, even though it may at times proceed in stops and starts because of fluctuations in the distribution of heat between atmosphere and ocean. Comparing this strong trend with CO_2 changes over the same time interval (Figure 1.1), it is easy to jump to conclusions. But let's avoid doing so: our purpose here is to develop an unbiased personal opinion. So we'll look at the fundamental workings of the climate system and deduce what may be happening, why this may be the case, what we might consider to do about it, and whether nature possesses any mechanisms that could help us with the solution.

Given that paleoclimate reconstructions are not something everyone encounters every day, I start with a chapter about the main techniques used in the discipline.

Next, we approach the issue of climate change from first principles, looking at Earth's energy balance (how much energy comes in versus how much goes out), which ultimately decides the surface temperature of the planet.

This is followed by a look at the main drivers of change in Earth's energy balance, from a natural perspective. We will see that greenhouse gases such as CO_2 and methane (CH_4) are unavoidably dominant in that discussion.

Then follows a discussion of why climate changes have occurred before human activity, how large these changes were, how fast the changes took place, and how the system recovered. This helps us understand how fast natural processes might "fix" any greenhouse gas and climate issues.

Next, we look at what we know about human influences. How do we know that today's greenhouse gas increases are caused by us, as opposed to being part of a natural cycle? How much will they affect climate? How fast is the human influence relative to natural cycles? Can nature cope with the human impact? In short, given an improved understanding of natural cycles, the question targeted is: *Does it seem likely that we're just in a natural cycle, and that nature can somehow come to the rescue and reverse the problem?*

At the end of every chapter will be a short section that recaps the key issues, as a basis for progression into the next chapter.

[2]

PAST CLIMATES
How We Get Our Data

On the work floor, research on past climates is known as paleoclimatology, and research on past oceans as paleoceanography. But they are very tightly related, and we shall discuss both combined under the one term of paleoclimatology. Within paleoclimatology, interests are spread over three fundamental fields.

The first field is concerned with dating ancient evidence and is referred to as chronological studies. These studies are essential because all records of past climate change need to be dated as accurately as possible to ensure that we know when the studied climate changes occurred, how fast they were, and whether changes seen in various components of the climate system happened at the same time or at different times.

The second field concerns observational studies, where the observations can be of different types. Some are direct measurements; for example, sunspot counts or temperature records. Some are historical, written accounts of anecdotal evidence, such as reports on the frequency of frozen rivers, floods, or droughts. Such records are very local and often subjective, so they are usually no good as primary evidence. But they can offer great support and validation to reconstructions from other tools.

Besides direct and anecdotal data, we encounter the dominant type of evidence used in the discipline. These are the so-called proxy data, or *proxies*. Proxies are indirect measures that approximate (hence the name *proxy*) changes in important climate-system variables, such as temperature, CO_2 concentrations, nutrient concentrations, and so on. This chapter outlines some of the most important proxies.

The third field in paleoclimatology concerns modeling. It employs numerical models for climate system simulation and simpler classes of so-called box-models. Numerical climate models range from Earth System models that are relatively crude and can therefore be set to run simulations of many thousands of years, to very complex and refined coupled models that are computationally very greedy and thus give simulations of great detail but only over short intervals of time. Box-models are much simpler and faster to run, and they are most used in modeling of the carbon cycle or other geochemical properties.

The different fields within paleoclimatology are complementary, and there is a lot of interaction: modeling results get grounded to real data, and interpretations of observations get checked with models for physical consistency. Often there are several such cycles involved from the first discovery of something to the presentation of more conclusive reconstructions. That way, the discipline gradually edges forward in its understanding. Through the last 50 to 60 years, the discipline has successfully identified the mechanisms behind the big lines of change in past climates. Work continues on improvements and refinements in detail.

The work is conducted according to an established tradition of formulation of initial hypotheses, followed by their critical testing and checking. The initial hypotheses are thus rejected or accepted. Then follow formulations of new hypotheses if the initial ones were rejected or the initial hypotheses get refined and updated. Then the cycle starts again. This is the normal process by which science advances in a careful, controlled, and self-critical manner.

In this book, I will not discuss modeling results very much. This should not be taken as an indication that modeling is not important. That conclusion would be a grave mistake. Modeling is essential for understanding how the system works, as models are physically and chemically consistent in a much more rigorous manner than interpretations of individual observational records. But models can also be very confusing to the nonspecialist because of the great variety of models that exists, all with different potentials and limitations. So I have chosen to make my case by using simple numbers only and referring to pertinent past (proxy) observations that are well established. In reality, many of the concepts and interpretations that I will discuss have arisen from—and have been refined by—model-based research working alongside observation-based research. The way I choose to approach things is simply to keep it transparent and flowing easily. Any further reading will immediately need to delve into modeling studies. Also, my emphasis on observational evidence (data and proxy data) addresses the often-voiced remark of *show me the data*.

In this chapter, I give an overview of the main methods by which we obtain data about past oceans and climate. The listing is not exhaustive, as there are many more important methods, but the main approaches are presented. I bypass historical measurements and written (anecdotal) evidence of climate change. After all, it is pretty obvious to all how a long time-series of temperature measurements by the admiralty, or by a meteorological service, might have been obtained and how it could be used to say something about historical climate states. Instead, I focus on proxy records for longer timescales because they shape our understanding of past climates in prehistoric times.

The most natural subdivision of proxy records for past climate change is by location (a) from ice; (b) from land; and (c) from the sea. I add a special section for reconstructions of sea level and global ice volume, for two reasons. First because ice volume is such a critical measure of the climate state, and second because I like it so much. It's what I work on every day.

2.1. DATA FROM ICE

The world's large continental ice sheets in Greenland and West and East Antarctica have existed long enough to give relatively undisturbed records that span very long periods of time. Cores drilled from the Greenland ice sheet have given high-quality, continuous climate records that extend back in time from the present day to almost 125 thousand years ago. For West Antarctica, such records go back a bit less than that. For East Antarctica, the records reach back to about 800 thousand or even 1 million years ago (Figure 2.1).

It had been known for a very long time from holes dug in snow and ice that there is a distinct layering, which results from seasonal cycles of snow accumulation. The weight of newly accumulating snow presses older layers into a much more compacted form, first called firn, and then ice. As pressure mounts deeper and deeper in the ice sheet, the annual layers are compressed thinner and thinner, until in the bottom portions the layers are too thin to be identified.

Where the layering can be recognized and counted in cores drilled through the entire thickness (several kilometers) of ice sheets, they give very accurate age scales to ice cores; a bit like counting tree-rings, but then counting downward from young to old. In the bottom portions of ice cores, where the annual layering cannot be recognized any longer, a different approach is needed. This is where ice modeling comes in. In the upper portions, ice models are tuned for the particular setting to the layering, and then they are applied further down to extend the age scale over the part where the layering is no longer recognizable. Dating of the lower portions as a result has larger uncertainties than dating in the interval where there is layering. There are many supporting techniques to this approach, such as the use of ash fragments captured in the ice from dated volcanic events, and the use of records of specific elemental isotopes that are formed by interaction of galactic rays with Earth's upper atmosphere. More information about these technical methods is best obtained from the appropriate specialist literature.[6]

In firn and snow, there are air pockets that still exchange with the atmosphere. But upon compression to ice, the air pockets become isolated and can no longer exchange with the atmosphere. Within the ice, therefore, there are bubbles that contain preserved ancient atmosphere, like miniature time-capsules. These bubbles can be analyzed to give detailed data of past atmospheric composition changes, including past CO_2 and CH_4 levels (Figure 2.1).[7,8,9,10]

Meanwhile, the ice itself can also be analyzed, and three of its properties have special importance. The first two are known as the oxygen isotope ratio and hydrogen isotope ratio of the frozen water molecules. Isotopes of any element are distinct in that they have the same number of protons in their core, but different numbers of neutrons. Molecules (such as H_2O) made of different isotopes therefore have slightly different masses. Despite the tiny differences in mass, molecules with somewhat lighter isotopes are a bit more reactive than those with somewhat heavier

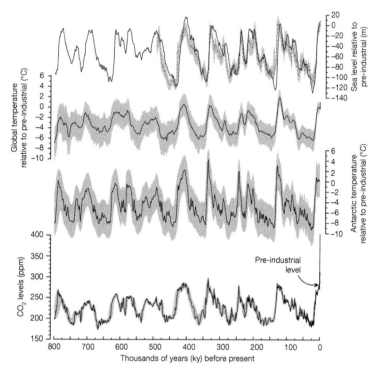

Figure 2.1. Key climate indicators for the past 800,000 years. Bottom: CO_2 fluctuations during the ice-age cycles of the last 800 thousand years, as measured in air-bubbles trapped in ice cores from Antarctica, and the extremely fast rise to 400 ppm since the start of the industrial revolution.[i,ii,iii,iv,v,vi] First panel up: Antarctic temperature changes relative to pre-industrial.[vii,viii] Second panel up: Global temperature changes relative to pre-industrial.[ix] Top panel: sea-level changes relative to pre-industrial (dashed from my own group's Red Sea method;[x] solid based on deep-sea isotopes).[xi] Some age differences can exist between records as a result of different dating methods. Where grey-shaded bands are given, these represent 95% confidence intervals; in other words they show how well we know the signal. Some studies fail to report them, as in the top panel.

i. Monnin, E., Indermühle, A., Dällenbach, A., Flückiger, J., Stauffer, B., Stocker, T.F., Raynaud, D., and Barnola, J.M., Atmospheric CO_2 concentrations over the last glacial termination. Science, 291, 112–114, 2001.

Figure 2.1. *Continued*

ii. Monnin, E., Steig, E.J., Siegenthaler, U., Kawamura, K., Schwander, J., Stauffer, B., Stocker, T.F., Morse, D.L., Barnola, J.M., Bellier, B., Raynaud, D., and Fischer, H., Evidence for substantial accumulation rate variability in Antarctica during the Holocene, through synchronization of CO_2 in the Taylor Dome, Dome C and DML ice cores. Earth and Planetary Science Letters, 224, 45–54, 2004.

iii. Schmitt, J., Schneider, R., Elsig, J., Leuenberger, D., Lourantou, A., Chappellaz, J., Köhler, P., Joos, F., Stocker, T.F., Leuenberger, M., and Fischer, H., Carbon isotope constraints on the deglacial CO_2 rise from ice cores. Science, 336, 711–714, 2012.

iv. Schneider, R., Schmitt, J., Koehler, P., Joos, F., and Fischer, H., A reconstruction of atmospheric carbon dioxide and its stable carbon isotopic composition from the penultimate glacial maximum to the last glacial inception. Climate of the Past, 9, 2507–2523, 2013.

v. Landais, A., Dreyfus, G., Capron, E., Jouzel, J., Masson-Delmotte, V., Roche, D.M., Prié, F., Caillon, N., Chappellaz, J., Leuenberger, M., and Lourantou, A., Two-phase change in CO_2, Antarctic temperature and global climate during Termination II. Nature Geoscience, 6, 1062–1065, 2013.

vi. Ahn, J., and Brook, E.J., Siple Dome ice reveals two modes of millennial CO_2 change during the last ice age. Nature Communications, 5, 3732, doi:10.1038/ncomms4723, 2014.

vii. Stenni, B., Masson-Delmotte, V., Selmo, E., Oerter, H., Meyer, H., Röthlisberger, R., Jouzel, J., Cattani, O., Falourd, S., Fischer, H., and Hoffmann, G., The deuterium excess records of EPICA Dome C and Dronning Maud Land ice cores (East Antarctica). Quaternary Science Reviews, 29, 146–159, 2010.

viii. Parrenin, F., Masson-Delmotte, V., Köhler, P., Raynaud, D., Paillard, D., Schwander, J., Barbante, C., Landais, A., Wegner, A., and Jouzel, J., Synchronous change of atmospheric CO_2 and Antarctic temperature during the last deglacial warming. Science, 339, 1060–1063, 2013.

ix. Snyder, C.W. Evolution of global temperature over the past two million years. Nature, 538, 226–228, 2016.

x. Grant, K.M., Rohling, E.J., Bronk Ramsey, C., Cheng, H., Edwards, R.L., Florindo, F., Heslop, D., Marra, F., Roberts, A.P. Tamisiea, M.E., and Williams, F., Sea-level variability over five glacial cycles. Nature Communications, 5, 5076, doi: x.1038/ncomms6076, 2014.

xi. Spratt, R.M., and Lisiecki, L.E., A Late Pleistocene sea level stack. Climate of the Past, 12, 1079–1092, 2016.

isotopes. Due to such effects during condensation and freezing of cloud vapor to form snow (ice), the oxygen and hydrogen isotope changes are well-understood recorders of the temperature at which the snow formed. As a result, time-series of oxygen and hydrogen isotope analyses from ice cores give us detailed insight into changes in the polar temperatures through time (Figure 2.1).

A critical third measurement from ice cores concerns wind-blown dust data. These can be obtained in different ways, by looking at either pre-served dust particles or dissolved dust ions in the ice. In ice ages, the world was more barren and drier (less vegetation), and there were increased levels of dust blowing around in the atmosphere. These dust veils are im-portant to climate, because they represent natural aerosols, which affect the radiative energy balance of climate. In addition, the iron in the dust veils fertilized the ocean, leading to increased algal production that helped lower CO_2 levels when the dead algal matter sank into the deep-sea.

There are many additional interesting and relevant proxy data from ice cores, but we don't need them here. Ice-core studies offer data in fine de-tail, and with good age control, and thus have developed into some of the most important sources of information about climate changes during the last 800 thousand years. There is no well-preserved ice from more ancient times, unfortunately.

2.2. DATA FROM LAND

First among land-based, or terrestrial, records for studying past climate change are stalagmites from caves in limestone regions, which are made of calcium carbonate. The dripwater from which they develop comes from percolation of rainwater through cracks in the limestone overhead. The main data obtained from cave deposits concern datings of when the deposits were growing (there was water) and when they were not (the cave was dry), and oxygen isotope ratios.

Dripwater oxygen isotope ratios, which are recorded in the calcite of the stalagmite, are affected by two main influences. First, they relate to the oxygen isotope ratio of the waters that were evaporated to fuel the rainfall—the so-called source waters. Second, they reflect the change of oxygen isotope ratio in the water vapor between the time it was first evap-orated and when it formed the rainfall. The latter change is influenced in quantitatively predictable ways by temperature changes in the atmosphere (especially at mid to high latitudes) and by the severity of the rainfall events (known as the amount effect; especially at low to mid latitudes).

The dripwater also transports dissolved carbon, and the carbon isotope ratio in that, which again is recorded in the calcium carbonate of the stalagmite, is a measure of the carbon in the soil system through which the rainfall has percolated. This is affected by vegetation changes. Thus, the carbon isotope values also help reconstruct rainfall amounts and temperature, but in a complex and not so easily quantifiable manner.

Stalagmite calcium carbonate can be very precisely dated using radioactive decay of elements in the so-called uranium decay series.[11,12,13] Such datings and the proxy measurements are often undertaken in great numbers down the center along the length of stalagmites. Thus, well-dated records can be made of well-understood proxies that tell us a great deal about climate changes over land. Cave deposits have contributed most information for periods younger than about 250,000 years ago, but improvements in dating methods are beginning to push that further back. Very much older stalagmites can be, and have been, investigated as well, but these cannot be dated precisely.

Tree rings offer another much-used terrestrial archive for paleoclimate information. These beautifully distinct sets of rings become visible when a tree trunk is cut through, or a drill core is taken from the outside to the center of the trunk. They are seasonal growth rings, with the youngest on the outside, and the oldest in the center. Counting the rings gives a detailed estimate of the tree's age. By using trees of different ages (they can be dated with the radiocarbon method), and looking at ring-thickness patterns, combinations can be made of tree-ring counts and thickness variations from many individual trees of the same species. Thus, composite series have been constructed that go back continuously to about 11,500 years ago, using data from both living and fossil trees.

The strict yearly age control obtained from counting tree-rings has been used to check and refine the radiocarbon dating method. This has greatly advanced the precision of datings from the very widely used radiocarbon, or ^{14}C, dating method. Besides great age control, tree-ring studies have also yielded much information about regional climate conditions, because tree-ring width and density are measures of regional temperature and rainfall conditions.[14] Thus, exquisitely well-dated records have been made of well-understood proxies that tell us a great deal about climate changes over land, back to about 11,500 years ago, but especially for the last 2000 years.

Sediment cores from lakes play an important role in paleoclimate studies as well. The sedimentary layers on the bottom of many lakes around the world have been drilled with a variety of equipment (and under a variety of challenging conditions—one of my colleagues once told me an interesting story about a Nile crocodile getting amorous or otherwise excited about a little motor-dinghy . . . while the researchers were in it). Sediments build up by gradual deposition from in-wash, biological remains, dust blown in by the winds, settling of volcanic ash, and so on. Especially lakes from higher latitudes often display fine laminations in the sediment, which reflect seasonal cycles of deposition. Like tree rings, these laminations, known as varves, can be counted downward to get very precise age control.[15,16]

In addition to varve counts, radiocarbon dating is often used to get age control, but it only works in the most recent 50 thousand years, and additional age control is obtained by using volcanic ash deposits from chemically identifiable eruptions. Thus, many lake records have very strong age control, although it can be quite weak in others. When studying lakes, it pays to search for one with varved sediments. A special case among "lakes" is the Dead Sea, which is actually very salty—some ten times more salty than the ocean. There, special circumstances mean that highly detailed dating is possible with the radiometric uranium decay-series method, and this excellent age control has been combined with highly detailed studies of changes in the lake level to offer a climatologically valuable measure of changes in the regional net evaporation.[17,18,19]

Most lake records extend back in time for several tens of thousands of years, while occasional ones can extend back hundreds of thousands of years. The long records are often not varved and can have dating issues, but dated volcanic ash layers within the lake deposits provide scope to make improvements in that respect. There is also support from Earth magnetic polarity changes in the sediments, which can be related to dated volcanic rocks that record the same changes in Earth's magnetic field.

Lake cores are commonly subjected to a wide variety of analyses. These types of analyses include plant pollen and spore counts to study vegetation changes around the lake; stable isotope analyses of carbonates and organic matter in the sediments to study precipitation changes, temperature changes, and biological process changes; organic biomarker molecule characterizations to see what lived in the lake and to get temperature records; and a wealth of studies of fossils and microfossils to reconstruct life in the lake and the conditions that controlled it. In addition, there have been many studies that quantified and dated changes in lake levels, which are used as measures of net dry or wet conditions.

Further climate information can be obtained from tell-tale landforms. The best-known examples are glacial landforms, and in particular heaps and ridges of material that was bulldozed along by ice. These are known as moraines. End-moraines especially, which are deposited along the outer edges of ice sheets and glaciers, tell us much about climate. End moraines can be dated using the material contained within them or buried underneath them, and they have given fascinating insight into variations of the North American and European ice sheets during the ice ages.

Moraines are especially suitable for studying the last ice age because the last ice age occurred about 20 to 25 thousand years ago, so its evidence is datable with the radiocarbon method, and also because landforms of earlier ice ages have been largely overrun and destroyed by ice of the last ice age. This, in combination with many other lines of research, has revealed that North America was covered during the last ice age by an ice sheet that was larger than the current ice sheet on all of Antarctica, and that there was an ice sheet about a third of that size over northern Eurasia.

Other detailed information comes from mountain-glacier moraines, which have revealed that these glaciers came about 1000 meters, and in some places, more than 1500 meters further down the mountains during the last ice age than they do today.[20,21] This gives us precious information about the temperature change with altitude in the atmosphere, which is a strong indicator of the global climate state. Mountain glacier moraines have furthermore been used extensively to study climate variations within the current warm period (the Holocene) from 11,500 years ago to the present. Mountain glacier fluctuations are one of the paleoclimate proxies that show a still poorly understood relationship of Holocene climate variability with longer-term sunspot variations, such as the Little Ice Age sunspot minima.[22]

There are many more highly specialized terrestrial methods in paleoclimate research, discussion of which would go beyond the scope of this book. But I will list two of the most notable ones to wrap up this part about terrestrial records.

In settings where plant leaves, or their wax coatings, are well fossilized, researchers have investigated the so-called stomatal density, which measures how many stomata are seen for each square centimeter of leaf surface. Stomata are like pores in the leaves, and the plant uses them to regulate its gas and vapor exchange with the atmosphere. Based on studies of modern plants and laboratory experiments, it has been found that the stomatal density is strongly determined by atmospheric CO_2 concentrations. Thus, researchers have used fossil leave coatings from well-dated sediment sections to reconstruct past changes in CO_2.[23,24]

Another, relatively new technique uses leaf waxes to reconstruct precipitation-evaporation conditions. Leaf waxes are organic coatings that are highly resistant to degradation. Upon decay of a leaf, the leaf waxes can be blown around by the wind, or even transported by rivers, and accumulate in sediments on the bottom of lakes or of the sea. In simple terms, once the leaf-wax compounds have been isolated from the samples, their hydrogen isotope ratios can be analyzed. These ratios are a strong indication of the regional precipitation-evaporation conditions in the area where the leaf was growing.[25,26,27]

2.3. DATA FROM THE SEA

The oceans are a treasure-trove of information about past climate changes. The sea floor is covered with layer upon layer of sediment. By drilling down, we can uncover information of past conditions. At the top is the youngest material, and we reach older and older material as we drill deeper. Observations from marine sediment cores tell us much about past climates because the ocean is an integral component of the climate system.

In most places in the open ocean, sediments accumulate at a rate of about 1 to 4 centimeters per thousand years, but in exceptional sites of high accumulation, the sediments can accumulate up to a meter or more per thousand years. Normal coring equipment can recover between 10 and 30 m of core, and specialist coring equipment deployed from regular research vessels can reach about 70 m. And then there is a deep-sea drilling system, which extends the maximum sediment-core recovery to many hundreds of meters (down to 2 kilometers when including ocean crust): this system is the Deep-Sea Drilling Project (DSDP), which has been often renamed and is now called the International Ocean Discovery Program (IODP). These ocean drilling programs have delivered continuous cores from the ocean that extend many millions of years back in time, with the current record-oldest sediments being about 170 million years old.

Once we have a sediment core on board, we need to get to grips with the ages of the material. Within the last 50 thousand years, this is relatively easy, as we can we date these young ocean sediments with the radiocarbon method. For older intervals, it becomes more difficult. There, we use a variety of methods. First to be used are magnetic polarity changes in the sediments, as also discussed above for lake records. Second are known and dated evolutionary switches in marine fauna and flora, based on studies of fossils and microfossils within the sediment record. Third, we use cyclic variations in oxygen isotope records that have been initially dated by means of the other techniques. Fourth, we refine age control using well-quantified relationships of the oxygen isotope cycles with astronomical changes in Earth's climate, which will be discussed in the next chapter.

The fourth step is called "astronomical tuning" of the age scales.[28,29,30,31,32] The method has been developed and used on independent sets of assumptions in the open ocean where ice-volume assumptions are used and in the Mediterranean Sea where monsoon-variability assumptions are used. In both settings, the findings have been validated over the last few hundred thousand years by relationships between the oceanic records and radiometrically dated cave stalagmite records at the margins of the basins, and over longer periods by using dated volcanic ash falls within the marine sediment sequences. In addition, rigorous tests have been applied to timescales based on astronomical tuning. Over the past five or six decades, this method has been applied, tested, adjusted, tested, and so on; this has been done so many times that the best astronomically tuned age scales are now surprisingly accurate and have great confidence assigned to them. This is how we can know, for example, how fast a large carbon isotope event (CIE) developed at around 56 million years ago, and how fast its carbon got "cleaned up" afterward. For interest, this dating method is so accurate that we can confidently state that the CIE of 56 million years ago happened because of a very rapid initial carbon injection event into the climate system within a few thousand years, while recovery of the climate system took some 200,000 years.

Now we understand the time frame of the sediment core. Next, we want to characterize the environmental and climatic changes that are recorded within it. The first proxy to be commonly applied in marine paleoclimate research concerns stable oxygen isotope records. These are measured on carbonate microfossils and have grown into the backbone of research on marine sediment cores. Almost everything else is hung from that framework.

We have briefly encountered stable isotopes before, but here we need a little more detail. The technique concerns measurement of the ratios of the stable isotopes of oxygen (^{16}O and ^{18}O) in marine microfossil shells. Molecules made up of isotopes of different masses display slightly different chemical reaction rates, and these rates in turn are sensitive to temperature. Thus, the ratio of oxygen isotopes in microfossil shells is a function of past temperatures, but it also reflects past oxygen isotope ratios of the seawater that the now fossilized organism lived in. In turn, past seawater oxygen isotope ratios are a function of global ice volume, the precipitation-evaporation balance over the water when it was in touch with the atmosphere, and mixing between different water masses with different oxygen isotope ratios.[33] It is not easy to disentangle these different influences, but intensive work by the research community has resolved most of this problem, by using parallel, independent measurements of some of the influences with other proxies, so that their effects can be subtracted. Notably, seawater temperature influences can be removed by using magnesium:calcium ratios, as will be discussed later. The remaining oxygen isotope signal then is easier to understand.

Oxygen isotope ratios can be measured on microfossils of organisms that used to live at the sea surface or on the sea floor. All material, including that derived from surface waters, will upon death sink to the sea floor to become part of the sediment. Thus, we can extract, from a single sample, different microfossils that used to live at about the same time at different depths. This approach allows us to make time-equivalent measurements for different depths in the ocean; for example, we can approximate profiles of temperature change with depth.

Microfossil oxygen isotope changes for the deep ocean are dominated by changes in global ice volume and deep-sea temperature. The latter can also be measured with Mg/Ca ratios of the same shells used for oxygen isotope analysis, so the temperature component can be removed from the oxygen isotope variations. This leaves a record that is dominated by global ice-volume changes—in other words, by ice-age cycles. These ice-volume changes can be translated into global sea-level variability (Figure 2.1).

Microfossil oxygen isotope changes for surface waters are a more complex mixture of signals from ice-volume changes, local precipitation-evaporation changes, and surface water temperature changes. The temperature component can again be removed using Mg/Ca-based temperature data from the same microfossil shells or using temperature data based on organic geochemical evidence from the same sample. The ice-volume component can be removed by using data from deep-sea microfossils in the same samples (see earlier in this section) or using other sea-level reconstructions. This then leaves the component that relates to local precipitation-evaporation changes, which is very useful in assessing past climate change. The inferences thus made about the local balance between precipitation and evaporation can be checked with independent data, for example from cave stalagmites on islands in the same region, or from new techniques such as hydrogen isotope ratios in plant leaf waxes from such an area. When working with proxy data, it is always important to check and validate results against those from other, independent methods.

The ratio of the stable carbon isotopes ^{12}C and ^{13}C is measured at the same time as the stable oxygen isotope ratio, from exactly the same microfossils. The stable carbon isotope ratio is a very complex mixture of signals,[33] and it goes beyond the scope of this text to delve into the matter. The only times when this ratio will be mentioned or used in this book, is in cases where the stable carbon isotope ratios throughout the hydrosphere-biosphere-atmosphere system shift to lower values (stronger dominance of ^{12}C) because of injection of so-called external carbon into that system from decomposition of methane gas hydrates, or combustion of fossil fuels. Here, external carbon refers to carbon that used to be safely locked away from interaction with the climate system, and which then was made available; be it by some geological event, or—as today—by human fossil-fuel burning.

Reconstructing temperature changes is key to understanding climate changes. For deep-sea temperature reconstruction, there is only one major proxy method. It employs the magnesium:calcium ratio (Mg/Ca) in carbonate microfossils. The great advantage of the Mg/Ca ratio from carbonate microfossils is that it is measured on exactly the same shells as the stable oxygen isotope ratio, so that it is then straightforward to make corrections to the latter based on temperature changes from Mg/Ca, as was discussed earlier.

Mg occurs as an impurity in $CaCO_3$ (calcium carbonate) formed in the ocean, and the degree to which Mg replaces Ca in the carbonates is temperature dependent. This temperature dependence most likely results from a temperature influence on species-specific biological processes in the organisms, given that the Mg/Ca to temperature relationship is different for different organisms studied. In any case, we can use Mg/Ca ratios to evaluate temperature changes if we always analyze the same species. To do so, the Mg/Ca of that species is calibrated to true temperature by using gradients in the modern ocean; we call this a core-top calibration. Typical uncertainties are about ±1°C.

For surface-water temperature reconstructions, there are more options. One is Mg/Ca temperature reconstruction using fossil carbonate shells of plankton microfossils. Another technique measures sea-surface temperature changes with organic geochemical compounds called alkenones, which originate from tiny single-celled algae called coccolithophores. Chemical differences in these organic compounds can be expressed in a ratio that is known to vary with temperature and that can be quantified by comparison with instrumentally measured, modern temperatures. The method can then be used to translate the organic geochemical ratios measured in sample sequences from dated sediment cores into a time series of sea-surface temperature changes.

There are other methods to evaluate temperature changes as well, such as further organic geochemical ratios and fossil assemblage variations, but the Mg/Ca and alkenone methods have become the dominant approaches. Of course, further development of the other proxies remains highly desirable, because in the end it is better to have many different ways of approaching a problem, rather than having to rely on just one or two. Indeed, this is true for all proxies: we are always trying to find additional, independent methods with which to compare results, enrich the evidence base, and build up confidence in the reconstructions that are made.

This brings us to a couple of proxies that have been developed to study CO_2 variations, which are needed to extend the ice-core-based CO_2 record beyond its limit of about 800,000 years ago (Figure 2.1). Over the last decade or so, the use of boron isotope ratios in microfossils has been taking flight as one of the most promising proxies for reconstruction of ocean pH, the chemical measure of acidity. This important measure can be used to calculate dissolved CO_2 levels in the surface ocean, and these in turn can be used to calculate CO_2 levels in the atmosphere.

The method uses fossil carbonate shells of plankton microfossils in samples obtained from sea-floor sediment cores. Studies concentrate on regions where the oceanic and atmospheric CO_2 levels are today observed to be close to equilibrium; that is, the levels of dissolved CO_2 in seawater are related directly to atmospheric CO_2 levels, with the least complications. First, variations in the boron isotope ratios in sample sequences through the sediment cores are used to reconstruct changes in ocean pH. Next, a series of calculations takes the reconstructed ocean pH to calculate the dissolved CO_2 level and that is, in turn, related to the atmospheric CO_2 level using Henry's Law of gas exchange.

There are other oceanic CO_2 proxies as well, notably one that relies on carbon isotope ratios in alkenones. These marine data can be compared with those from other methods, such as one that uses carbon isotope ratios in sodium-carbonate minerals (notably nahcolite, which is a fancy name for baking soda), the leaf stomatal density method that we saw earlier, and in the most recent 800,000 years, of course, the ice-core CO_2 measurements.

Overall, the various techniques agree about the "big-picture" CO_2 changes over millions of years, but much work remains to sort out the finer details. In consequence, uncertainties become quite large in deeper time, but a decent long-term record can still be reconstructed (Figure 2.2).

Based especially on modeling work and careful ocean carbon-inventory reconstructions, the well-documented CO_2 variations associated with ice-age cycles (Figure 2.1) have been attributed mainly to carbon exchange between the ocean and the atmosphere.[34,35] But a more direct observational basis for CO_2 storage variations in the oceans had been missing. While boron isotopes in oceanic plankton microfossils helped us measure upper-ocean and atmospheric CO_2 variations, we still needed something by which to measure fluctuations through time in the amount of carbon storage in the massive deep-sea volume. One could do this by using boron isotope ratios in carbonate microfossils of unicellular bottom-dwelling organisms known as benthic foraminifera, but there are strong analytical challenges that prevent this for now from becoming a major technique. In contrast, promising results have been obtained over the last decade from work on boron:calcium (B/Ca) ratios in benthic foraminifera, which agree well with results from boron isotope ratios in studies that considered overlapping data.[34]

B/Ca ratios in benthic foraminifera have been related to carbonate ion concentrations in the seawater in which they live, by means of core-top calibrations. Calculations then relate changes in the carbonate ion concentration in seawater, as derived from B/Ca changes, to dissolution of CO_2 in that seawater. Changes in the dissolution of CO_2 in the ocean relate to CO_2 uptake from, or release to, the atmosphere.

During ice-age cycles, atmospheric CO_2 variations predominantly resulted from redistributions of carbon within the hydrosphere-biosphere-atmosphere system. Storage into, and release from, the vast deep-ocean reservoir were particularly important in those processes.[34,35] Over timescales of five to ten thousand years, intensive storage of dissolved CO_2 into the deep ocean starts to interact with carbonate sediments. It causes carbonate in the deep-ocean sediments to dissolve. The opposite is true when deep-ocean CO_2 is reduced, which over time will reduce the dissolution of sedimentary carbonate. The net result of this process, which we refer to as carbonate compensation, is that the amount of CO_2 storage in or

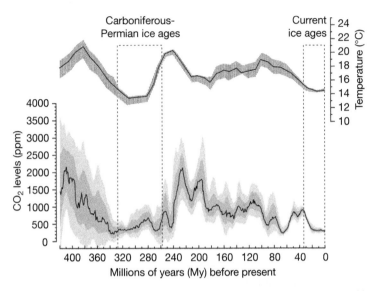

Figure 2.2. Key climate indicators for the past 420 million years. Bottom: Earth's long-term CO$_2$ fluctuations.[xii] Gray shading indicates 68% (dark gray) and 95% (light gray) confidence intervals. Top: Earth's long-term temperature fluctuations.[xiii] In the temperature graph, the uncertainty shading indicates the difference between two model settings, and not a specific level of confidence. Interestingly, we now find ourselves in the curious position that we have a better understanding of past CO$_2$ fluctuations than of past temperature fluctuations. Also indicated (boxes) are the two main periods characterized by ice ages. The portrayed fluctuations are only the long-term changes. Superimposed, there have been shorter time-scale variations. Note that the two compilations are from different sources, and that apparent age-offsets between them must be considered with care—these would need to be examined in detail before conclusions may be drawn. Where this has been resolved, the coincidence between CO$_2$ and temperature changes is remarkable (Figure 2.1).

xii. Foster, G.L., Royer, D.L., and Lunt, D.J. Future climate forcing potentially without precedent in the last 420 million years. Nature Communications, 8, 14845, doi: 10.1038/ncomms14845, 2017.

xiii. Royer, D.L., Berner, R.A., Montanez, I.P., Tabor, N.J., and Beerling, D.J. CO$_2$ as a primary driver of Phanerozoic climate. GSA Today, 14, 4–10, 2004.

release from the deep ocean, for changes over thousands of years or longer, is larger than what one would estimate by just measuring the changes in deep-ocean CO_2 values. This is because carbonate compensation accounts for a large proportion of the change over those long timescales.

Similarly, when so-called external carbon is injected into the hydrosphere-biosphere-atmosphere system, the deep-ocean CO_2 levels will rise, and this will drive carbonate dissolution. Thus, the deep ocean will extract a disproportionate amount of carbon. During the Paleocene-Eocene Thermal Maximum, which we will encounter in section 4.1, a large venting of carbon into the ocean-atmosphere system (partly from previously isolated stores deep with the sediments) triggered strong carbonate dissolution throughout the world ocean, which over hundreds of thousands of years removed the injected excess carbon from the ocean atmosphere again.[36,37]

We can use this carbonate compensation response to get a feel for long-term CO_2 changes in the past, going back many tens of millions of years to perhaps 100 million years ago, by which time evolution had established all essential, modern major players in the carbonate cycle of the oceans. In very simple terms: if strong dissolution developed, then CO_2 probably had been increased, and if dissolution became reduced, then CO_2 had probably been decreased. We can measure this, using movements over time in something called the carbonate compensation depth (CCD), which effectively is the depth below which carbonate deposition does not occur. CCD movements over time can be reconstructed using marine sediment cores from different water depths.[38,39] Note that any use of the carbonate compensation process in this way for detailed reconstructions will require serious modeling.

2.4. DATA ABOUT SEA-LEVEL CHANGES

At this stage, it is important to go into some detail about how we reconstruct sea-level changes, given that this is one of the most explicit measures of change in the Earth system, including climate. I will first look at methods that we have to reconstruct sea-level changes on timescales of centuries to a few million years. But for understanding Earth's very long-term cycles (Figure 2.2), we need to be able to reconstruct sea level over millions to hundreds of millions of years, and I will conclude with a short explanation of how we can approximate such changes.

On timescales of a few centuries to thousands of years, global continental ice-volume change is the key determinant for global sea-level change: the more continental ice volume, the lower the sea level (Figure 2.2). Today, the world still contains enough global ice volume to cause a rise in sea level of about 65 meters if it all melted. Sea-level change, or global ice-volume change, can be reconstructed in several ways. In the following discussion, I focus on the main techniques that are used over periods of more than 10,000 years. The first uses fossil corals, which are radiometrically dated via the radiocarbon method or, more frequently, uranium-series dating. The second method uses deep-sea stable oxygen isotope data from sea-floor microfossils (benthic foraminifera), with correction for temperature changes from Mg/Ca analyses of the same material. The third method uses records from marginal seas that have only a narrow and shallow connection with the open ocean, such as the Red Sea and Mediterranean Sea. Finally, there are some other methods that I will summarize before launching into very long-term sea-level cycles.

Most corals live close to the sea surface because they require light to support their photosynthesizing symbionts. Symbiosis is an arrangement whereby two organisms gain significant benefits from living in a combined manner—one inside the other. In corals, the symbionts that live within the coral polyps are single-celled algae, which practise photosynthesis. They gain protection and concentration from this arrangement, and the coral polyp gains an additional energy/food source. Note that symbiosis is different from parasitism, because in that case one organism lives at the expense of another. There are also corals that live in the dark deep-sea, without symbionts, but then they gain their energy by actively feeding. In shallow settings, corals both actively feed and gain energy from their photosynthetic symbionts. Because light is needed for the photosynthesis, large coral colonies (reefs) build up in shallow tropical waters.

By registering which coral species live at certain depths, a sense has been built up of the relationship between different coral associations and water depth. This knowledge can be used with fossil coral reefs from drill cores or from uplifted reef terraces. In addition, these fossil corals can be dated accurately by radiometric techniques, such as radiocarbon for the last 50,000 years, and the Uranium-series for the last 250,000 years, and with possible extension back to 500,000 years ago. So now we have dated samples of corals whose preferred living depth can be obtained from biological studies. A bunch of additional measures is needed, especially for uplift or subsidence through time of the area where we found our coral sample, but in essence this is what the coral sea-level method gives us: a dated point (or series of points) of a coral species with known depth preference below sea level.

By building up many such datapoints around the world, we get a sense of how sea level has varied through time. It is a method with much appeal, because it relies on organisms that everyone can picture in relation to sea level. But there are enough uncertainties in this game to require independent sea-level methods to validate and complement the information from corals. Also, dating of corals becomes problematic for times before about 250,000 years ago, so other sea-level methods are essential for older times.

Previously, we discussed oxygen isotope records from the deep sea and how these can be used to calculate past ice-volume (hence, sea-level) changes, preferably using a deep-sea temperature correction based on Mg/Ca data. This is a widely used method, and one of the most detailed continuous records yet extends all the way back to 1.5 million years ago.[40] In principle, this method might be employed to look all the way back to the earliest onset of ice build-up on the planet at around 40 million years ago in Antarctica,[41] although the Mg/Ca applicability becomes questionable in older times, and thus needs more work. Before that time, and ever since the ending at around 250 million years ago of the previous main period of ice ages, there has been no significant land ice on the planet. In that period of about 250 to 40 million years ago, deep-sea oxygen isotope records hold no information about ice volume—because there was none—and instead were almost entirely dominated by deep-sea temperature changes.

All proxy methods that one could devise will always hinge on some assumptions and uncertainties. So the aim is to apply and compare several different methods, where these need to be independent from each other in the sense that they do not rely on the same assumptions and uncertainties. To support and test the reconstruction from the aforementioned sea-level methods, my own research group has developed such an independent approach, known as the marginal-seas method of sea-level calculation (Figure 2.1, see note 10).[42,43,44,45]

The method arose from observations that stable oxygen isotope changes associated with the more recent ice-age cycles were strongly amplified in the Red Sea and Mediterranean, relative to such changes over the same cycles in the open ocean. It was quickly realized that this was caused by the relative isolation of these seas and their limited communication (water-mass exchange) with the open ocean. This exchange is limited because very shallow passages connect the basins with the open ocean. The passage is only 137 m deep in the southern Red Sea Bab-el-Mandab strait, and only 284 m deep in the western Mediterranean Strait of Gibraltar. Compare those depths with the sea-level changes from about 120 to 130 m below the present during ice ages, to present-day like values in the intervening warm periods, known as interglacials.

Only so much water can be exchanged through a small sea strait. It's a bit like a garden hose. If you pinch it, first the flow gets more intense, or faster, but if you pinch it more, the flow will reduce and eventually stop. It only presses its way out because the pressure inside the hose builds up very strongly, while outside the nozzle the air pressure is still low (1 atmosphere). In a sea strait, the pressure contrast can build up a bit but not nearly as strongly as in our hose-example. Part closure of the strait therefore rapidly leads to reduction of the water-exchange through the strait. This effect can be accurately calculated using hydraulic control models.

A reduction of water exchange with the open ocean, because of a lowering of sea level, causes water in the Red Sea and in the Mediterranean to be exposed longer to the high evaporation over those regions. This affects both the salt content and oxygen isotope ratios of the seawater, which can be measured in microfossils. Changes in the oxygen isotope ratios of microfossils from the Red Sea and Mediterranean can then be related to sea-level changes via the hydraulic control models for water-exchange through the straits. Sea-level records from this method have been constructed back to 500,000 years ago in the Red Sea, limited by the maximum length of sediment cores from that basin. From the Mediterranean, much longer cores are available, and the record extends all the way back to 5.3 million years ago.[42]

The Red Sea and Mediterranean sea-level records can be compared with Mg/Ca-corrected deep-sea oxygen isotope records. This reveals good agreement in the reconstructed ice-volume variations from the present back to about 1.5 million years ago, and reasonable agreement before that time.[42] Much of the current work is focused on improving the records for times before 1.5 million years ago.

Several other methods are being used to reconstruct sea-level changes over one or a few ice-age cycles.[46] One uses the microfaunal zonation in salt marshes right in the tidal zone. This zonation of microfauna is documented in detail in modern saltmarshes and then applied on sediment cores taken from the same salt-marsh region. Dating is commonly provided by radiocarbon analyses. This way, very accurate records of regional relative sea level can be developed, especially for the current interglacial period, the Holocene, since the end of the last ice age.

A second method has found considerable use in gently sloping margins, such as the Sunda Shelf, Indonesia, and Bonaparte Gulf in northern Australia. This method takes cores along transects from the modern coast and into deeper waters, and then identifies when the sea level rise after the last ice age drowned the coastline. The drowned land-plants can be dated with the radiocarbon method, and the depth of the flooding surface noted from careful measurement. Thus, well-dated records are obtained of coastal drowning at different depths below the present sea level, with some corrections for uplift or subsidence of the land. This is then taken as a dated record of the sea-level rise after the end of the last ice age.

A third method to consider sea-level rise is similar to the land-drowning method, except that it uses coastal caves that at times emerged because the sea level fell and at other times drowned because the sea level rose. Stalagmites in the cave grow when the cave has emerged above sea level and stop growing when the cave is drowned. These changes can be dated with the uranium-series method, and using the elevation of the cave we can thus get well-dated information of when the level of the cave was crossed by sea-level change. These represent important point markers of past sea levels.

The sea-level methods discussed to this point are all applicable on timescales of centuries to a few million years. But we also need to know about the Earth's climate state in much more ancient times. To know about sea-level changes millions to hundreds of millions of years ago, a method is used that was developed in close collaboration with the petroleum industry. The industry was interested in sea-level cycles because these caused packages of deposits in continental shelf and slope settings that in some cases form potentially valuable petroleum reservoirs. The method is extensively explained elsewhere,[47] and I will summarize only the main concepts here.

The first principles of this method were worked out by a team of ESSO geologists who produced the "Vail curve" or "Exxon curve" in 1977, and it extends almost 550 million years into the past. A significant update came in 1987–1988, which is now known as the "Haq curve." A recent update has placed the Haq curve on an upgraded timescale.[48]

The most surprising finding, for most people, is that sea levels were up to several hundreds of meters above the present over much of the last 550 million years. This seems contradictory with the knowledge that there is at most 70 meters worth of ice on the planet today, and that moving all the freshwater in the world (including groundwater) into the oceans would not even double that number. To get sea level well over 100 m higher than today, we clearly need to think of other processes. It turns out that these processes are related to the large-scale splitting up, movement, and re-combination of continental plates on Earth, which are collectively known as plate tectonics.

During periods of intensified plate tectonics, the ocean crust is relatively young and buoyant; its mid-ocean spreading ridges are hot and bulging, and new sections of underwater ridge are formed as well. By displacing water, this reduces the volumetric capacity of the oceans, which in turn causes sea level to rise even if there was no change in the total volume of seawater. The long-term rises and falls in sea level related to plate-tectonic activity develop very slowly, but the process over time leads to large-amplitude changes. We have nothing to worry about from these processes, as humanity probably will not be around on Earth long enough to see these very slow processes make an impression on sea level—we are talking about tens of millions of years. Some of the shorter (and smaller-amplitude) sequences mapped out in the Vail and Haq curves are faster and relate to processes discussed earlier, such as ice-volume variations.

2.5. RECAP AND OUTLOOK

In this chapter, we have gone in broad strokes through some of the main lines of observational evidence used in paleoclimatology. It is understood that modeling has played an equally large role in developing the knowledge we now possess, but in the interest of brevity and in answer to calls of "show me the data," emphasis was placed on the observational (data and proxy data) aspect of the paleoclimate knowledge base.

We encountered data from ice cores, which go back to about 800,000 years ago and include "time-capsules" of fossil atmosphere in the form of air bubbles trapped in the ice, as well as isotopic ratios of the ice itself and both ion-series and discrete particle data on atmospheric dust fluxes. We next reviewed data from land, ranging from cave deposits to tree rings, and from lake deposits to landforms and plant leaf-wax data. This then brought us to discussion of data from the sea. We discussed sediment coring and dating, including the important astronomical time-scale tuning, and then a variety of methods that rely upon microfossil shell-chemistry. Special attention was given to proxy data that underpin reconstructions of past CO_2 concentrations, and we finally went through the various methods by which past sea levels can be reconstructed.

We are now ready to address the question: *What caused the past climate changes, before human influences?* First we look at some critical fundamental background in chapter 3, and then we address the actual question in chapter 4.

[3]

ENERGY BALANCE
OF CLIMATE

The Sun is the ultimate energy source for climate. The Sun radiates toward Earth at an almost constant intensity of about 1360 watts per square meter (W/m^2), as measured above the Earth's atmosphere. Most of this radiation takes place in the short ultra-violet and visible light wavelengths. We refer to it as incoming short-wave radiation (ISWR; the wavelengths are short because the Sun radiates at very high temperatures of about 5500°C).

Earth is not a two-dimensional disk, but a 3-dimensional sphere. Its day-side faces the Sun and receives radiation, while its night-side is directed away from the Sun and does not receive solar radiation. As a result, the global average energy received from the Sun per square meter of Earth surface is the energy received by the day-side of Earth averaged over the surface area of the entire sphere. When we do the mathematics, this gives an average input of solar radiation into every square meter of Earth, at the top of the atmosphere, of 340 W/m^2 (Box 3.1). That is the value that things work out to when considering the ISWR from the Sun in a continuous and globally equally "smeared out" sense, and that is what matters when we are working out the balance between energy gained and lost by Earth (Box 3.2).

Many people are puzzled by the fact that we talk only about energy from the Sun. They then especially wonder why we ignore heat input from the deep Earth, and in particular from volcanoes, which after all are very hot. But in spite of the spectacular shows of heat, steam, gases, and primordial mayhem that volcanoes put on display, they turn out to be almost

Box 3.1.

Earth intercepts the 1360 W/m^2 of incoming solar radiation according to its cross-sectional area, which is given as the area of a disk: πr^2. However, the surface of Earth over which we calculate average incoming radiation is the surface of a sphere: $4\pi r^2$. As a result, the average incoming radiation distributed over each square meter of Earth's surface is $1360/4 = 340 \text{ W/m}^2$.

Box 3.2.

It is useful to form a real-life feeling for the intensity represented by the 340 W/m^2 of global average energy gain from the Sun at the top of the atmosphere. For a first example it means that, on average, Earth continuously receives enough energy to power almost 6 bright (60 W) light bulbs within every single square meter of its surface. And then bear in mind that there are 510 trillion of those square meters, where a trillion is a million millions. Another useful number for comparison is that an adult person constantly loses about 100 W of heat. So to sense how much energy continuously comes in from the Sun (global average), imagine an experiment as follows. Find a small room, such as a big closet or a small toilet room, with a surface area of 2 m x 1 m (2 m^2). Then find 6 adults, step in so that you add your own 100 W, and close the door to stop any heat exchange with the outside. It will quickly get very unpleasant in there, as anyone will tell you who has ever been stuck in a full elevator. This is the sort of heat input we get on average from the Sun at the top of the atmosphere, all day, every day.

negligible in terms of heat flow into the climate system. Compared with the global average solar energy gain of 340 W/m^2, recent assessments show that total heat outflow from the Earth's interior is not even 0.09 W/m^2.[49, 50] About half of this comes from primordial heat related to the planet's formation, and the other half from radioactive decay of elements inside Earth.[50] And yes, these total heat flow numbers include volcanoes. This comparison puts into perspective what the Sun does for us. Clearly, heat loss from Earth's interior does not make an impression on climate, relative to heat received from the Sun, on a globally averaged scale. So when considering global climate changes, we can ignore any direct influence of heat flow from the Earth's interior.

Here it must be stressed that, when we later come to consider very long-term climate cycles over hundreds of millions of years, heat flow from the interior of Earth returns into the discussion with an important indirect role. On those timescales, the interior heat flow drives Earth's plate tectonics—the slow movement of continental plates around the world—that dominate these very long-term climate cycles.

Another common question is why we are so interested in the top of the atmosphere. This is simply because that is the boundary between Earth and space, where we have to evaluate the balance between Earth's loss of energy to space, and its gain from the Sun. This balance is known as the Earth's energy balance. After warming up due to absorption of ISWR (similar to any object warming up by absorbing sunlight), Earth cools by radiating heat out into space, similar to any warm body. Because Earth's surface is not very hot, most of the outgoing radiation occurs at the relatively long wavelengths of thermal infrared. Technically, we know it as outgoing long-wave radiation, or OLWR. It is basically the same as that used in images from night-goggles or thermal infrared cameras. You may have seen in such images that the intensity of OLWR depends on the temperature of the object that radiates the heat, so the rate of cooling of Earth by OLWR depends on Earth's temperature. More on OLWR will be discussed later. Here, all we need is the essence of Earth's energy balance: if Earth gains more energy from radiation at the top of the atmosphere than it loses, then it warms up; if Earth loses more energy than it gains, then it cools down.

To understand how much energy Earth actually gains from the Sun, we need to look at what happens with the ISWR in some more detail (Figure 3.1). Not all the ISWR that reaches the top of the atmosphere actually manages to become absorbed by the Earth surface, and it is absorption of radiation that is needed to cause warming. Instead, a considerable portion is reflected straight back into space without effect. The shiny white surface of ice or snow reflects more than 80% of the ISWR that hits it. Specific types of bright, white clouds are also excellent reflectors. Bare desert sand does a good job at reflection too—you'd be well advised to protect your eyes when travelling through the desert, just as in snow. When most radiation is reflected straight back into space, little is absorbed, which is why materials with reflective surfaces do not warm up very efficiently. In contrast, open, deep water reflects very little and absorbs most of the radiation that hits it, especially in low latitudes where the Sun rises to an almost directly overhead position. Dark vegetation and soil also absorb a lot of ISWR. Such strongly absorbing materials heat up efficiently.

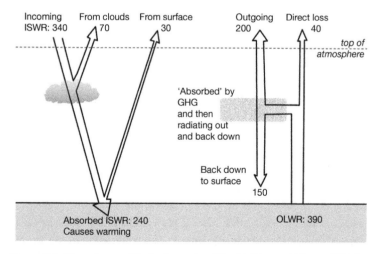

Figure 3.1. Strongly simplified schematic of Earth's Energy Balance. ISWR is incoming short-wave radiation. OLWR is outgoing long-wave radiation. GHG stands for greenhouse gas. All values are given in W/m².

We represent reflectivity by a factor named albedo. Today, Earth's global average albedo is about 0.3, mostly as a result of clouds, snow, and ice. This number means that 30% of the 340 W/m^2 of ISWR received at the top of the atmosphere is reflected back into space. Only about 70% of it, or 240 W/m^2, is absorbed at the Earth surface, where it causes heating.

The reflectivity of Earth is not constant through time. If we ignore clouds for now and look only at surface albedo changes, then we can get an interesting sense of the scale of change by looking at an ice age. During the last ice-age maximum, about 30,000 to 20,000 years ago, North America was covered by an ice sheet of similar size as that seen today over Antarctica, and northern Europe and northwest Asia hosted an ice sheet about a third of that size. The presence of such big, reflective ice sheets reduced the global average absorbed ISWR during the ice age by up to 4 W/m^2 from the modern value of about 240 W/m^2. Meanwhile, other reflectivity changes added a reduction by a further 4 W/m^2 or so, due to sea-ice expansion, vegetation changes, and increased atmospheric dust levels under drier and windier conditions.[51] Thus, the absorbed ISWR component was reduced to roughly 232 W/m^2 during the ice age. The reflected component was about 340 − 232 = 108 W/m^2, so global albedo was about 108/340 = 0.32 during the ice age, relative to about 0.3 today, in response to the surface changes.

It's worthwhile to allow ourselves to get a little side-tracked here. The ice-age albedo calculation illustrates how important it is to be careful with making decisions based on apparently small numbers, if one doesn't have a sound understanding what they stand for. An albedo increase from 0.3 to 0.32 might seem very small. I can almost hear a comment: *It's only 2 percentage-points of change! How important could that be?* But in reality, this tiny change represents a completely different-looking planet. It has no space for any land-based life throughout most of North America and northern Europe, where the land is covered by kilometers-thick ice sheets; additional thick and permanent ice over Alaska and other mountainous regions around the world; major permafrost extension through most of north Asia and into central Europe, with associated elimination of forests, which gave way to tundra; a vast sea-ice cap around Antarctica, which at times almost doubled the apparent size of that continent; another vast sea-ice cap that completely clogged the Arctic and at times extended as far as south of Iceland; and a global sea level that stood some 120 to 130 m lower than today, which drastically reshaped the outlines of the continents. So dismissing effects because of "tiny numbers" may be very misguided in view of the real impacts on the ground.

But let's return to the main flow of our argument. We will need to get a bit technical about feedback, a critical characteristic of responses within the climate system to an initial disturbance. Feedback is where an initial change or disturbance triggers a response that amplifies (positive feedback) or dampens (negative feedback) the impact of that initial change or disturbance. Most of us know an example of positive feedback from squealing microphone systems and of negative feedback from the dampening impact of a shock absorber.

Changes in reflectivity (albedo) drive a powerful cascade of such feedback effects: when temperature is lowered for whatever reason, vegetation reduces, and ice and snow cover increases. This increases reflectivity, which reduces ISWR absorption, and this in turn causes further cooling of the surface. This is known as a positive feedback. There is then a knock-on effect, which will be discussed in more detail in the next section: as the surface cools, OLWR intensity reduces, and that limits further cooling of the surface. This is a negative feedback. The end result is a shift of the Earth system toward a new, lower equilibrium temperature, at which reduced ISWR absorption and reduced OLWR emission are balanced once again. These processes can work in the opposite direction too, causing the Earth system to shift toward a new, higher equilibrium temperature at which increased ISWR absorption and increased OLWR emission are balanced once again.

The "pathways" toward the end-results that are followed by feedback processes are not simple, because the many processes that are involved all operate over different timescales (Figure 3.2). Growth and decay of large ice sheets on land is a slow affair that takes many thousands of years, while vegetation changes take decades to thousands of years. In contrast, sea-ice and snow-cover responses and dust effects are fast and operate within years. Thus, we distinguish slow albedo feedbacks (land-ice and vegetation related) from fast albedo feedbacks (clouds, water vapor, sea-ice, snow, and dust related). Researchers have been working for decades on understanding the major controls on, and timescales of, the changes

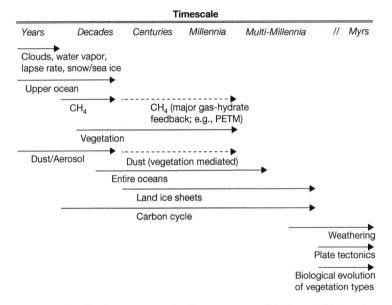

Figure 3.2. Feedback processes in the climate system and their typical timescales of operation. Myrs stands for millions of years; PETM stands for Paleocene-Eocene Thermal Maximum, a warm spike in Earth's climate history at around 56 million years ago; and // indicates a break in the timescale axis between timescales of several thousand years and timescales of millions of years.

Modified from http://www.realclimate.org/index.php/archives/2015/03/climate-sensitivity-week/

in such processes, to improve their representation in climate models. Especially atmospheric dust influences are difficult to understand and to represent in detail, as is the balance between albedo and OLWR-blocking characteristics of different types of clouds.

Now we have come to a point where we can consider what happens when Earth's surface is warmed up. At absolute zero, about −273°C, an object in space cannot lose any heat. As soon as an object is warmer than that, it will lose heat to space. As mentioned before, the intensity of Earth's outgoing long-wave radiation (OLWR) depends on its temperature: when Earth warms up, it will start to lose more energy through OLWR, and when

Earth cools down, its energy loss through OLWR will go down. This way, changes in the OLWR intensity help keep Earth's temperature from running away. In technical terms: the Earth system seeks a new equilibrium at which the increased/decreased net outgoing radiation once again equals the increased/decreased net incoming radiation.

Without the atmosphere, Earth's temperature would be around -18°C, based on the Stefan-Boltzmann Law, but the observed Earth surface temperature is some 33 degrees centigrade higher than that, at around +15°C. Clearly, something acts like insulation, so that Earth has warmed up to a point where (even with the insulation in place) the increased intensity of OLWR allows sufficient energy loss into space to balance the ISWR. The insulation is the atmosphere, and its insulating capacity is known as the greenhouse effect. Even before humans started to affect the atmosphere, the greenhouse effect was well established. It is the very reason that water exists on Earth in liquid form, which makes the planet's surface conditions favorable for life. Life and the greenhouse effect are close allies.[52]

If we apply the Stefan-Boltzmann Law to calculate OLWR at the actual observed average global surface temperature of Earth, then we find an OLWR of 390 W/m^2 from the Earth surface (Figure 3.1). That seems odd, because net (after reflection) absorbed ISWR amounts to only 240 W/m^2. So given that Earth has not been on a mad cooling dash toward a permanent and complete ice-ball state, a very-near-to-equilibrium state must exist in which net OWLR from Earth into space equals net ISWR from the Sun. In plain terms, this means that some 390 − 240 = 150 W/m^2 of the OLWR generated at Earth's surface must be retained in the atmosphere, simply because it clearly is not escaping into space.

We have now seen two ways of looking at the greenhouse effect: one is in terms of temperature (33°C), and the other is in terms of energy that is retained by the atmosphere (150 W/m^2). So let's have a look at what causes it.

3.1. THE GREENHOUSE GASES

We start with a quick history of research into the greenhouse effect. This is essential to demonstrate that we're not just looking at an untested theory, but instead at a thoroughly tested and validated scientific understanding. The earliest elements of research hark back to more than 150 years ago.

We know which components of the atmosphere are responsible for the greenhouse effect by studying which wavelengths of the OLWR are partly blocked from escaping into space, due to absorption within the atmosphere. The OLWR spans a wide range of wavelengths. Humans can't see in the infrared range, so it's easier to visualize a shorter-wavelength example: think about the wide range or spectrum of different colors of light in a rainbow. Each color represents light of a different wavelength, from violet and blue at shorter wavelengths to red at longer wavelengths. Different gas molecules absorb radiation at different wavelengths. Next, these molecules re-emit the absorbed energy from cooler levels higher up in the atmosphere, both back down to the surface and upward into space. The resultant weakening of emission into space at specific wavelengths can be measured with so-called spectrometers, which then show a "dip" or "well" in the spectrum of radiation at the blocked wavelengths. To follow our visible-light example, there would be dark bands along the rainbow, where certain colors are weakened.

Theory and experiments have determined which molecules cause spectral gaps at specific wavelengths of radiation. Already in the 1860s, John Tyndall established experimentally that several gases in the atmosphere, including CO_2, absorb radiant heat.[53] Earth's outgoing radiation has been studied with spectrometers that are mounted on satellites and high-altitude aircraft. Thus we know that on Earth, partial blocking of OLWR due to absorption within the atmosphere—the greenhouse effect—is mainly caused by water vapor, while CO_2 and methane (CH_4) are also important. There are other contributors, like ozone, but I focus only on the three major ones here: water vapor, CO_2, and CH_4. Spectral absorption calculations indicate that about half of the total 150 W/m^2 of the greenhouse effect is caused by water vapor, roughly a fifth by CO_2, and the remainder by CH_4 and other components.

Water vapor is our most important greenhouse gas. It is also rather special. This is because water vapor in the atmosphere is controlled almost exclusively by temperature, through what is known as the Clausius-Clapeyron relationship. The higher the temperature, the more vapor the atmosphere can hold. Because water vapor concentrations change rapidly with temperature, water vapor is not a "driving" greenhouse gas, but instead a "following" greenhouse gas that provides a critically important positive feedback to changes in climate. If temperature goes up for whatever reason, then the atmospheric water vapor content goes up. This causes more blocking of OLWR by absorption in the atmosphere, which causes warming, and this in turn causes an increase in intensity of OLWR from the surface. This continues until a new equilibrium is reached between net loss of OLWR and net gain of ISWR at the top of the atmosphere.

It is important to emphasize that, in the absence of the other greenhouse gases, all water would quickly condense and fall out of the atmosphere. The other greenhouse gases "drive" the greenhouse effect, and water vapor merely amplifies their impact as a fast-acting feedback in the climate system. Note that, in addition to its OLWR-blocking impacts, increases in water vapor also affect cloud cover. Some clouds are highly reflective and can cause reduction in absorbed ISWR.

CO_2 and CH_4 are different because they cannot condense out of the atmosphere. Both are much more rare in the atmosphere than water vapor. CO_2 is measured in parts per million (ppm), by volume. In the natural warm climate states before human activity that occurred in between the ice ages of the last one million years, the atmosphere held about 280 ppm of CO_2, or 0.028 per cent (Figure 2.1). CH_4 is even more rare; it is measured in parts per billion (ppb). In the same natural warm times in between ice ages, CH_4 levels were about 700 ppb, or 0.00007 per cent. Despite their scarcity, these gases are so effective at partly blocking OWLR that they cause clearly measurable spectral gaps. As early as 1896, Svante Arrhenius published a seminal study[54] that described the greenhouse effect due to CO_2.

Because CO_2 and CH_4 are rare in the atmosphere, intensive human-driven (anthropogenic) processes such as fossil fuel burning and cement production (Figure 1.2), along with deforestation and agricultural activity, have been able to cause important CO_2 and CH_4 increases from natural levels of 280 ppm and about 700 ppb, to more than 400 ppm and 1800 ppb, respectively. We are clearly re-shaping the natural concentrations of these powerful greenhouse gases. The last time CO_2 levels were anywhere near today's level of about 400 ppm was in the Pliocene, some 3 million years ago. With no industrialized humans around, such past CO_2 levels must be explained by natural causes.

Under natural conditions without human influences, CO_2 and CH_4 levels varied due to processes in the natural carbon cycle. The carbon cycle spans the complex web of interactions that controls carbon storage and exchange between the biosphere (life), hydrosphere (oceans, lakes, rivers), and lithosphere (rocks and sediments). Its processes play over timescales that range over anything from days to months, years, decades, and centuries, and that even extend to many thousands or millions of years where lithospheric influences are concerned. The carbon cycle will be discussed in depth in chapter 4.

For any given disturbance, some parts of the carbon cycle may respond quickly (e.g., plant life), but most of its major processes respond slowly. For example, carbon exchange between the atmosphere and deep ocean takes thousands of years, partly because it takes a long time to circulate the deep ocean, and partly because of a process called carbonate compensation, as explained in the next chapter. Lithospheric processes of sediment and rock formation, weathering, and volcanic changes are even slower, and take millions of years. As a result, CO_2 and CH_4 levels are never in complete equilibrium with climate—they are always moving along some delayed pathway toward a new equilibrium. Because of their long adjustment timescales, carbon cycle feedbacks to climate changes are classed as slow feedbacks. Their tardy responses can lead to severe unbalances between fast changes in the climate state and slow ongoing adjustments of CO_2 and CH_4.

The really interesting, if somewhat confusing, upshot of this all is that changes in CO_2 and CH_4 levels may both be the driver of change (for example during a long episode of intense large-scale volcanicity, commonly related to continental plate movements), and be major feedbacks through the various carbon cycle components to any initial change in climate. So if we see CO_2 changing with climate, this does not have to mean that CO_2 was the ultimate trigger of the change, but it certainly will be one of the major feedbacks to affect the change once it has started. In turn, the impacts of changes in CO_2 and other greenhouse gases will be strongly amplified by the fast-acting water-vapor feedback. Similarly, if CO_2 does change first—as today under anthropogenic release of this gas—then the temperature change that it causes cascades through the various feedback processes as well, with several amplifying and some weakening effects. In such an intricate web of feedbacks, it doesn't really matter much where we start.

Because the carbon cycle has many slow components, complete CO_2 adjustments may sometimes seem to lag behind the initial climate changes by hundreds to thousands of years. This does not mean that CO_2 "only follows climate change," or that "CO_2 does not drive climate change," as we sometimes hear. It only means that CO_2 is part of a critical feedback mechanism to any change, whatever its cause. It is a bit like a bicycle without freewheeling capacity: if you push its pedals, its wheels will turn. But also, if its wheels start to turn (perhaps because it's on a slope), its pedals will rotate. It does not matter where in this feedback cycle you start to push. In the end, all components will turn faster, and you can make everything turn faster in different ways; for example by directing the bike down a steeper slope, or pushing harder on its pedals, or both. If you want to include an analogy for the effect of slow carbon cycle responses, then think about connecting the bicycle's pedals with the wheels by means of a stretchy chain. At first, the stretch accommodates for all the change, giving a response time lag between pedals and wheels. But eventually all components will be turning together again. It's important to keep this complex interlinked nature of the system in mind. If not, then we risk making bad mistakes in searching for simple cause-and-effect relationships.

3.2. A PERSPECTIVE FROM STUDIES OF PAST CLIMATES

Paleoclimate studies have revealed that, over the past hundreds of millions of years of geological time, long and warm periods with high CO_2 levels up to a few thousand ppm alternated with colder periods of generally low CO_2 levels, reaching as far down as 100 to 150 ppm. These long cycles of large-scale CO_2 change (Figure 2.2) resulted from movements of continental plates and associated mountain-formation, weathering, volcanism, and other lithospheric processes.[52] We are currently in a low-CO_2 phase of that long-term variability, with CO_2 levels only up to 280 ppm in the natural warm states of the past one million years, and down to 180 ppm during the ice ages within that period (Figure 2.1). These generally low CO_2 levels explain why frequent ice ages occurred within the past million years, and why large ice sheets survived even during the intervening warmer intervals—in the present-day warm phase, after all, there are still very large ice sheets on Antarctica and Greenland, which together hold enough water to cause some 65 m of global sea-level rise if they all melted.

If we take a long-term perspective and look at the past 500 million years, then we find that it is rather exceptional to have much ice on Earth (the presence of large ice masses is easy to detect in geological evidence). More than 90% of that time, Earth was warmer than today. Through the Triassic to Cretaceous Periods (252 to 66 million years ago), when dinosaurs roamed the Earth, tropical oceans at times reached almost 35°C and polar regions some 15 to 20°C. The pertinent question is: Why was Earth so warm?

To get a rough sense for what was going on, we don't need highly sophisticated models. Simple arguments and sums suffice. With no ice sheets, and lush vegetation extending from tropics to poles, Earth would be less reflective than it is today. In technical terms: it would have a lower albedo. But it is hard to imagine how this would increase the absorbed ISWR by more than about 10 W/m². In addition, the continents and oceans were differently distributed than today, and the locations, orientations, and heights of mountain ranges were different[52]. Those influences may have caused an additional change in the absorbed proportion of ISWR. Let's make a generous assumption that this doubled the excess of absorbed ISWR, relative to the present, from 10 to 20 W/m². We then get a rough estimate that absorbed ISWR may have been about 260 W/m².

For a global average temperature of some 10 degrees centigrade higher than today—that is, about 25°C—we can use the Stefan-Boltzmann Law to find that OLWR from the surface would be almost 450 W/m². Net OLWR into space must have virtually balanced net ISWR after reflection, or else climate would have gone into runaway heating or cooling, cooking or freezing the dinosaurs, which did not happen. So net OLWR into space must have been close to 260 W/m². In consequence, we find that 450–260 = 190 W/m² of OLWR from the surface never made it into space. In other words, 190 W/m^2 was retained by the atmosphere. Today, the value of retained OLWR is 150 W/m^2. This simple sum indicates that, in the warm period of the dinosaurs, the greenhouse effect appears to have been some 40 W/m^2 more intense than today.

Now we should step back a bit and consider whether this rough evaluation might possibly be completely wrong, for example because the Sun was stronger in the past, which would mean an increase in the 260 W/m^2 assumed previously. The simple answer is no, because evolution of a Sun-like star implies that the Sun became more intense with time, over hundreds of millions to billions of years. Over the last billion years, the Sun's so-called luminosity has increased by about 7 to 8%,[55] or roughly 0.7 to 0.8 % per hundred million years. So during the warm era of the dinosaurs, the Sun would have been 0.5 to 2.0% less intense than today. For sure, our earlier evaluation is not wrong because of a brighter Sun; if anything, the Sun was a little bit less bright. With no extra incoming energy, the explanation for the excess warmth of the era of dinosaurs therefore must be sought in the insulating role of the greenhouse effect.

This then brings us to a next question, namely: *How much of the extra greenhouse effect might be explained by an increase in the atmospheric water-vapor content in the warmer world?* For a global average temperature change from 15 to 25°C, the Clausius-Clapeyron equation indicates that the so-called saturation vapor pressure in the atmosphere will be increased by roughly 90%.[56] In "plain speak," this means that the total amount of water vapor that could be held in the atmosphere will be almost doubled. Relative humidity is measured against that saturation vapor pressure; it is the ratio of how much vapor there is, relative to the maximum there could be. If we assume that relative humidity was not changed, then a 10°C global average temperature increase implies that water vapor would account for roughly 20 of the 40 W/m^2 additional greenhouse effect during the warm period of the dinosaurs.[57] So the water-vapor feedback explains about half of the extra OWLR retention in the atmosphere of the dinosaurs' warm period. Therefore, the other half must have resulted from other greenhouse gases. Note that a similar 50:50 split has been determined from spectral absorption studies for the modern greenhouse effect. Put simply, when increases in greenhouse gases like CO_2 and CH_4 cause an increase in atmospheric

OWLR retention, and therefore temperature, we can expect that direct impact to be roughly doubled by the water-vapor feedback.

If the remaining excess 20 W/m² greenhouse effect of the past warm period relative to the present resulted entirely from CO_2, then we can determine the past CO_2 concentrations. This is because every doubling of CO_2 levels (for example, from 140 to 280 ppm, from 280 to 560 ppm, or from 560 to 1120 ppm) is known to cause almost 4 W/m² of change. An increase of 20 W/m² therefore means five doublings of CO_2 levels relative to those before the industrial revolution, which were around 280 ppm. Five doublings would go from 280 to 560 to 1120 to 2240 to 4480, and finally to 8960 ppm. Thus, we obtain a high estimate of almost 9000 ppm for the warm era of the dinosaurs. But this is an extreme estimate, in which CO_2 is held responsible for everything except the water-vapor component and other greenhouse gases are completely ignored. If we instead assume that CO_2 accounted for about a fifth of the total 40 W/m² change, similar to its contribution to the modern greenhouse effect, then CO_2 accounted for 8 W/m². That is equivalent to two doublings in the CO_2 concentrations relative to the pre-industrial value. We thus obtain a rough lower-end estimate of 1120 ppm for CO_2 levels in the warm world of the dinosaurs. This lower-end estimate agrees well with estimates made from geological measurements (Box 3.3; see also Figure 2.2).[58]

Our two scenarios allow us to make a rough estimate of a critical term in climate science, namely Earth System Sensitivity. This is the amount of total warming relative to a change in CO_2 concentrations, with the latter expressed in terms of the number of doublings of the CO_2 concentration from the pre-industrial 280 ppm. In the extreme case where we calculated five doublings of CO_2, we infer an Earth System Sensitivity of 10°C change per 5 doublings, or 2°C change per doubling. In the more realistic scenario, we infer an Earth System Sensitivity of 10°C per 2 doublings, or 5°C per doubling. How realistic is this crudely estimated range of 2 to 5°C per CO_2 doubling? Well, a census of a wide variety of studies of modern and past climates concluded a likely range of 2.2 to 4.8 °C per CO_2 doubling.[59] And a detailed assessment for the climate between 2.3 and 3.3 million years ago, when CO_2 most recently was at similar levels to today's, has recently

Box 3.3.

Because we don't have any methods to measure past CH_4 levels—or those of the other, minor, greenhouse gases—in deep geological time, our simple arguments could only limit our estimates of CO_2 levels in the warm world of the dinosaurs to somewhere between 1100 ppm and 9000 ppm. Comparison with geological measurements suggests that the lower end of our range is more realistic. This indicates that the proportionalities of the greenhouse impacts of water vapor, CO_2, and the other components appear to have been rather similar to those observed today (before the industrial revolution, that is); about 50% was due to water vapor, about 20% due to CO_2, and the remainder due to other components. Such "stability" of these proportionalities hints at the existence of an intricately linked natural response cycle through closely interlinked feedback processes within the carbon cycle and the hydrological (water) cycle.

confirmed that.[60] Our crude example may not be so bad after all; it may be rough, but it'll do for the sake of illustration.

3.3. RECAP AND OUTLOOK

In this chapter, we have discussed the Energy Balance of climate. We determined the global annual mean Incoming Short-Wave Radiation (ISWR) received from the Sun at the top of the atmosphere, and how different reflection (albedo) factors reduce it before part of the ISWR becomes absorbed by the Earth's surface, where it causes warming. We argued that, overall, this net absorbed ISWR must be very closely balanced by Outgoing Long-Wave Radiation (OLWR) at the top of the atmosphere, because Earth is not in a runaway ice-house or greenhouse state. But based on surface temperature observations, we saw that the OLWR lost from the Earth's surface is some 150 W/m^2 higher than the OLWR going into space. This is the energy retained by the atmosphere; it is the greenhouse effect in terms of energy. Along the way, we also determined the greenhouse effect in terms of temperature, and we found that Earth is about 33°C warmer than it would be without an atmosphere.

As part of these discussions, we looked at the main feedback processes in climate and their timescales. And we also encountered the main greenhouse gases. In terms of importance, we saw that water vapor is the main one (about 50%), but it is a "condensing" greenhouse gas that only follows and amplifies temperature variations driven by the other greenhouse gases. Next in line of importance is CO_2 (about 20%), while CH_4 and other gases make up the rest. We considered humanity's impacts on carbon emissions, driving a fast and large greenhouse gas rise since the onset of the industrial revolution, and we started to compare this with natural variations of the past 800 thousand and even 500 million years.

In the end, we found that the greenhouse effect offers a sound explanation for past warm periods. But humans and their ancestors (for example, *Homo habilis, H. erectus, H. ergaster, H. heidelbergensis, H. neanderthalensis,* and *H. sapiens*) have been around only for about 2.5 million years, of which *H. sapiens* spans a mere 300,000 years[61], and our total hominin lineage with a mostly upright posture only seems to extend back to about 4.2 million years ago. Obviously, therefore, humans had nothing to do with the climate changes of the distant past, yet still there were large changes in the greenhouse gas concentrations (Figures 2.1, 2.2). This confuses many people.

Three main questions are commonly raised:

1. With no humans emitting greenhouse gases, what made their concentrations go up and down?
2. With natural cycles of rises and drops in greenhouse gases, how do we know that the current greenhouse gas increase is not just part of such a natural cycle?
3. Accepting that major natural CO_2 variations happened before humans were around, nature clearly has ways of dealing with CO_2—so won't nature simply "come to the rescue" and remove the human-emitted greenhouse gases from the atmosphere?

To answer the first question, chapter 4 looks at the drivers of natural climate variations before human intervention, including the natural processes that used to govern greenhouse gas concentrations. The second question is evaluated in chapter 5 by comparing modern changes with the past (natural) changes. The third question is addressed in chapter 6, drawing on observed limits to the rates of natural climate-change processes.

[4]

CAUSES OF CLIMATE CHANGE

The causes of natural climate variations, before human impacts, typically arose from one or more of the following: carbon-cycle changes, astronomical changes in the Earth-Sun configuration, large volcanic eruptions (especially plate-tectonics-related major volcanic episodes), asteroid impacts, or variations in the intensity of solar radiation output. Carbon-cycle changes may have acted on their own but were often also involved as a feedback in amplifying the climate responses to changes driven initially by the other mechanisms.

In the following sections, we will look at each of these processes in turn.

4.1. CARBON-CYCLE CHANGES

When we want to discuss the dominant changes in greenhouse gas concentrations, we focus mainly on CO_2 and CH_4, of which CO_2 is the dominant one on longer timescales because it exists in much higher concentrations and lasts much longer in the atmosphere than CH_4. As mentioned before, we then commonly investigate things in terms of carbon (C) emissions and uptake because this allows us to relate variations directly to changes in the carbon cycle and how we humans are affecting it.

The carbon cycle represents an intricate web of interactions that control carbon storage and exchange between the biosphere (life), hydrosphere (oceans, lakes, rivers), and lithosphere (rocks and sediments;

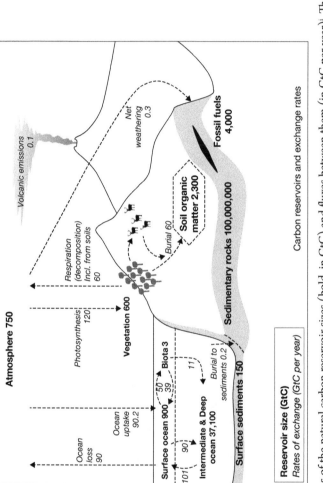

Figure 4.1. Approximations of the natural carbon reservoir sizes (bold, in GtC) and fluxes between them (in GtC per year)[i]. The values given for the fluxes are roughly after those determined for the natural system of the past 1 million years (before humans)[ii]. There is a net unbalance of carbon removal (sediment burial + weathering) versus carbon input (volcanic emissions) of roughly 0.4 GtC per year. This natural unbalance would cause a gradual loss of carbon from the atmosphere-ocean-biosphere system that might eventually translate into a very slow CO_2 drop. Human actions, however, have caused up to 10 GtC per year additional input (Figure 2) on top of this natural system, causing a fast CO_2 rise.

i. Rohling, E.J, The oceans: a deep history. Princeton University Press, 272, 2017. ISBN9781400088665.

ii. Wallmann, K., and Aloisi, G. "The global carbon cycle: geological processes." In Fundamentals of Geobiology, ed. Knoll, A.H, Canfield, D.E., and Konhauser, K.O. First Edition, 20–35. Blackwell Publishing Ltd, 2012. http://geosci.uchicago.edu/~kite/doc/Fundamentals_of_Geobiology_Chapter_3.pdf

Figure 4.1). We need to consider two critical terms when discussing it. The first is known as the reservoir volume. This stands for the volume of carbon held within each reservoir, such as land-plants and trees, the ocean, or carbonate rocks. The second term is known as flux, and it refers to the amount of carbon that is exchanged between two reservoirs in a year. Because the volumes of carbon that are involved are enormous, we commonly express them in gigatons (Gt). One gigaton is one billion (one thousand million) tons, where a ton is 1000 kg. Most frequently, this term will be used in the expression gigaton of Carbon, or GtC.

There are several important reservoirs of carbon (Figure 4.1). The atmosphere holds an approximate volume of 750 GtC. The land-biosphere—living flora and fauna—comprises some 600 or 700 GtC of living material and more than 2000 GtC of dead material. The ocean holds 38,000 GtC of dissolved inorganic carbon, the bulk of which is located in the deep ocean. Fossil fuels and shales combined contain more than 12,000 GtC, of which "pure" fossil fuels make up at least 4000 GtC. Marine surface sediments hold 150 GtC, and sedimentary rocks hold at least 100 million GtC. In contrast to all these major reservoirs, the ocean biosphere only contains 3 GtC of living material, and we will ignore it.

Carbon fluxes between the reservoirs are indicated in italics in Figure 4.1, together with their approximate values in GtC per year (GtC/y). Under natural circumstances—before humans had a notice-able impact on the carbon cycle—the fluxes were very closely matched in terms of net carbon exchanges between the sedimentary rocks (and the fossil fuels they contain) and the active climate-system components (the atmosphere, ocean, and biosphere). You may recognize these from section 2.3 as exchanges of so-called external carbon.

If the input and output of external carbon weren't very closely balanced, then there would have been a fast runaway to zero CO_2 or maximum CO_2 in the atmosphere, which did not happen on Earth: atmospheric CO_2 fluctuations have been contained between about 100 and a few thousand ppm (Figure 2.2). In contrast, a runaway to maximum CO_2 has happened on Venus, where CO_2 now makes up more than 96% of the dense atmos-phere, and where temperatures are close to 460 °C.

But it is important to note that the balance between external carbon input and output is not absolutely perfect. Most times, there have been tiny offsets in the natural system's net external carbon flux. Times with minor net carbon gains in the active climate system were marked by slow CO_2 increases, and times with net carbon loss from the active climate system caused slow CO_2 drops. Although tiny, these net external carbon flux offsets acted over unimaginably long geological timescales of millions to hundreds of millions of years, and this allowed their impacts to accumulate into long-term CO_2 variations over the observed range of a few thousand ppm[62] (Figure 2.2). During the past million years, there has been a minor net loss (Figure 4.1).

Even under natural circumstances, occasional inputs of external carbon have happened in a faster manner through major natural CH_4 and/or CO_2 "burps" that caused shorter and sharper climate events over thousands to a few hundred thousand years. We will see later in this section that the CO_2 change for a major example of such events is estimated between 500 and 3000 ppm, at rates of 0.3 to 1.7 GtC per year,[63] or at narrower estimates of 0.6 to 1.1 GtC per year.[64] By comparison, the human-caused "burp" of external carbon input into the climate system has in recent years grown to a massive 10 GtC per year[65] (Figure 1.2). The human impact modifies the natural balance of Figure 4.1 by addition of a flux-arrow between the fossil fuels reservoir and the atmosphere of up to 10 GtC/y. For the natural "burps" this extra arrow would have had a value of 0.3 to 1.7 GtC/y.

When considering change in the carbon cycle, we need to make a fundamental distinction between two key types of change. The first type of change in the carbon cycle concerns variations that involve net external carbon addition or removal from the hydrosphere-biosphere-atmosphere system. The second type concerns fluctuations in the carbon cycle that involve no net additions or removals, but only redistributions of carbon between the hydrosphere, biosphere, and atmosphere; that is, *within* the active climate system. We discuss each type in more detail next.

The first type of carbon cycle variation typically involves carbon exchange with the lithosphere—the rocks and sediments. For example, carbon removal occurs by formation of sediments that subsequently get buried and thus excluded from interactions between the hydrosphere, biosphere, and atmosphere. Or there is net carbon addition from rocks and sediments, or from fossil fuel reservoirs, which were previously not exchanging with the hydrosphere, biosphere, and atmosphere. This may happen through volcanic outgassing of CO_2 released by the "cooking" of sediments where marine plates with sediments are pushed underneath continental plates in subduction zones; think, for example—of the "ring of fire" around the Pacific. It may also happen through oxidation of organic matter that was held in sediments (often shales) or of "pure" fossil fuels (oil, gas, and coal) as they become exposed through cycles of mountain-formation and weathering.

Incidentally, oxidation of fossil fuels could be happening either naturally—which normally represents a small flux—or it could be "artificially assisted" through fossil-fuel mining and burning by humans. The impacts of large and fast increases of external carbon input of any kind are immediately obvious: CO_2 levels in both the atmosphere and ocean would need to respond with a correspondingly large and fast rise, causing a tandem of warming due to CO_2 rise in the atmosphere, and ocean acidification due to increased CO_2 invasion into the ocean.[52] Because a natural "burp" of external carbon into the climate system has similar consequences to a human-caused input of external carbon, we can look to records of such natural events in the geological record to get a sense of the potential scale and duration of the consequences of human action.

The big natural driver behind Earth's long-term, large-scale atmospheric CO_2 cycles is the intensity of Earth's so-called plate tectonics. As was mentioned previously, plate tectonics concerns the movement of continental plates around the world, in places drifting apart due to opening of new oceans by spreading at mid-ocean ridges (a strongly volcanic process), and in other places drifting together by destruction of ocean basins at so-called subduction margins (even more conducive of volcanism).[52] This process of eternal movement is ultimately driven by energy from the planet's interior heat engine.

Weathering of silicate and carbonate rocks forms the crucial link between volcanic outgassing of external carbon into the atmosphere and removal of carbon from the hydrosphere-biosphere-atmosphere system by formation of marine sediments. Weathering of these types of rock consumes atmospheric CO_2 and is intensified during periods with exposure of fresh rocks, notably resulting from the formation of new mountain ranges. The formation of new mountain ranges is in turn intensified during periods of active plate tectonics. So intensified plate tectonics cause both more volcanic outgassing that starts to build up atmospheric CO_2 and slow formation of new mountain ranges that through weathering begin to consume CO_2. In the form of bicarbonate (HCO_3^-) and carbonate (CO_3^{2-}) ions in river water, the carbon extracted by weathering makes its way into the ocean, where it eventually gets locked away into marine sediments. A complicating factor in the weathering influence is that weathering of different types of rock has different impacts on CO_2 levels. Weathering (oxidation) of sedimentary organic matter is a source of CO_2 to the climate system. Weathering of silicate rocks consumes more CO_2 per chemical unit (mole) of rock minerals than weathering of carbonate rocks. Full calculation of long-term CO_2 variations therefore needs to consider not only such processes as volcanic outgassing and temperature-dependent global weathering rates, but also changes in the world-averaged rock-type that's being weathered[66] because these changes affect the net amount of carbon consumed by weathering even if total weathering rates would remain constant (see note 2 of Figure 4.1).

All the processes involved in the long-term carbon variations act on different timescales. These differences then cause the tiny imperfections in the balance between external carbon input into, and removal from, the hydrosphere-biosphere-atmosphere system, and these in turn build up over the extremely long geological timescales involved into long-term highs and lows in atmospheric CO_2 levels (Figure 2.2). Again we witness the existence of an intricate web of interactions that helps keep CO_2 levels within certain limits.

It's only under highly exceptional circumstances that these intricate checks and balances may fail, which would lead to an irreversible runaway situation, such as on hothouse Venus (Box 4.1). Although plate tectonic processes are the big driver, life is the key thing on Earth that does the fine-tuning. It regulates carbon uptake and release via both organic and inorganic (carbonate) processes, and even regulates weathering rates through microbial and plant-root processes. It may not be unreasonable to assume that an absence of life on Venus played some role in things coming off the rails there, while its presence on Earth helped keep everything within reasonable limits.

Let's now have a closer look at the faster events ("burps") of external carbon input in the geological past, within a few thousands of years. Several of these events took place during the Paleocene and Eocene Epochs (about 66–34 million years ago) and are recognized in the form of striking anomalies in stable carbon isotope ratios—they have become known as carbon isotope events (CIEs). There are three types (isotopes) of carbon atom, which differ by the number of neutrons in their core: Carbon-12, Carbon-13, and Carbon-14 (radiocarbon). All have six protons in their core, which is what makes them carbon. ^{12}C has six neutrons, while ^{13}C has seven, and ^{14}C has eight. The higher the number, the heavier the atom. ^{12}C and ^{13}C are stable through time, but ^{14}C suffers radioactive decay. The ratio of the two stable carbon isotopes relative to each other, and relative to the ratio in a laboratory standard carbonate, tells us much about the source of the carbon that is registered. The stable carbon isotope ratio is very low in methane, and less low in marine sediments.

Through the CIEs, all reservoirs of carbon analyzed—be it land-based or ocean-based—show a considerable anomaly of low stable carbon isotope ratios. This indicates that external carbon must have been injected into the hydrosphere-biosphere-atmosphere system from a large source of material with a less low carbon isotope ratio (mostly CO_2), or a less large source of material with a very low ratio (methane), or some combination. Based on calculations along these lines, it has been estimated that the most conspicuous of the CIEs, the Paleocene-Eocene Thermal Maximum (PETM) of about 56 million years ago, coincided with a CO_2 increase of between 500 and 3000 ppm, and that the injection took place at rates of 0.3 to 1.7 GtC/y.[63] Measuring the carbon emission in terms of amount of CO_2 increase makes sense even if the original emission was in the form of methane, since methane very rapidly oxidizes to CO_2.

The PETM was one of the (if not the) largest and fastest climate events of the past 100 million years, or more. To put it into perspective, however, the human-caused external carbon emissions today are about 10 GtC/y, and the time-averaged human-caused emissions over the past century have been about 4 GtC/y (Figure 1.2). The time-averaged human-caused external carbon emissions therefore are 2 to 13 times faster than the PETM injection of external carbon, while the human-caused emissions in recent years have become some 6 to 33 times faster than the PETM injection.

Box 4.1.

Although major long-term climate fluctuations have taken place, they appear from geological evidence to have always been contained within a rather limited range. In other words, the long-term balancing processes on Earth's greenhouse-maintained climate conditions have proven to be very powerful. To appreciate this, we need to briefly consider the two possible runaway extremes: boiling or a deep freeze. Concerning the boiling extreme, radiation calculations indicate that there is a maximum rate of cooling from loss of OLWR of roughly 320 W/m^2 at temperatures of 100°C and relative humidity of 100% (steam coming from the oceans).[57] The planet would have to become almost totally non-reflective for net ISWR to balance that rate of cooling, which is unlikely. Therefore, we can safely state that, at least until the sun gains sufficiently in intensity at around a billion years from now, Earth will be safe from a runaway humid greenhouse state in which the oceans would boil off (by the end of which time, temperatures of almost 400°C will be reached). In fact, Earth has been closer to the "danger level" at the opposite, deep-freeze end of the scale. About 700 million years ago, Earth several times went through "snowball Earth" phases that saw it completely covered by firm or slushy ice.[52] At those times, the Sun was weaker by about 5 to 6%, and greenhouse gases dropped low enough for temperatures to fall below zero. As a result, all water condensed out of the atmosphere (Earth became "freeze dried"), so that the atmosphere could retain hardly any OLWR. But Earth still managed to break free from that stranglehold. It is thought that volcanic emissions related to Earth's plate tectonics, driven ultimately by the planet's interior heat engine, were crucial in ending these episodes. Earth thus remained within the "surface temperature zone" of 0 to 100°C at which liquid water can exist on the planet, throughout its lifespan of 4.6 billion years.

We still don't fully understand what exactly caused the carbon release of the PETM and other CIEs. Key suspects are CH_4 release from so-called gas-hydrates in ocean sediments, and CO_2 and possible CH_4 outgassing from pulses of intensified volcanism and other magmatic processes. Regardless of the underlying processes, the consequences of the carbon injection were dramatic. There was about 6°C of global warming within about 20,000 years, most of which happened within about 4,000 years. Though dramatic, this is still some 4 to 22 times slower than the roughly 1°C rise in about 150 years that has happened since the industrial revolution (Figure 1.1). The large and fast PETM injection of external carbon into the world ocean also caused lowering of the ocean pH, in a process known as ocean acidification[67,68], and this in turn caused strong dissolution of carbonates in the oceans[36, 37, 52].

After the initial injection, the excess carbon could only be removed by net loss from the hydrosphere-biosphere-atmosphere system, which is a very slow process, as we have seen. Given that the size of the PETM injection of external carbon is similar in magnitude to the human-made emissions projected in "business-as-usual" scenarios (although at a roughly 10 times slower rate), the PETM provides a useful estimate for the likely duration of a natural clean-up of our emissions. That duration was some 100,000 to 200,000 years[52]. This gives the game away: nature does not seem to have any mechanisms to clean up large CO_2 emissions faster than over a period of a great many thousands of years.

The second type of carbon cycle fluctuation involves no external carbon input or loss. Instead, it is all about redistributions of carbon between the hydrosphere, biosphere, and atmosphere, in response to changes caused by other drivers. In this type of change, atmospheric CO_2 levels depend purely on internal feedback processes within the carbon cycle. These variations typically lead to "short" CO_2 cycles that span "only" many hundreds to thousands of years. At some times, there were relatively fast CO_2 changes of about 100 ppm in just a few thousands of years (Figure 2.1).

On these relatively short timescales, the net offset between carbon burial and volcanic outgassing is so small that it remains negligible. Instead, carbon-cycle components mostly play a feedback role on these timescales, driving internal redistributions of carbon between the hydrosphere, biosphere, and atmosphere. One complication is that, over timescales of 5–10 thousand years, ocean sediments join the party through the process of carbonate compensation, in which carbonate can dissolve into the water from some places (especially the deep sea) and precipitate from the water in other places (notably shallow regions).[34,35]

Within the hydrosphere-biosphere-atmosphere system, the hydrosphere—especially the ocean—holds the most carbon (Figure 4.1). For example, the amount of carbon in the oceans is some 50 times larger than the total amount of carbon in the atmosphere. The terrestrial biosphere contains a similar amount of carbon as the atmosphere if we count only living biomass, and some three times that amount if we include also dead biomass. Therefore, any a minor change in the carbon content of the ocean through exchange with the other two reservoirs causes large changes in the atmosphere and terrestrial biosphere.

The carbon content of ocean water is strongly dependent on temperature because colder water can hold more dissolved inorganic carbon. The deep sea, below 1500 m depth, holds the largest quantity of carbon for a combination of that reason and because of decomposition at depth of sinking dead organic matter. It also helps that the deep sea represents about 60% of the world's ocean total volume. Exchange of deep-sea water with shallower layers, which is needed to connect deep-sea properties with the atmosphere, strongly depends on the degree of mixing in the ocean that is associated with its large-scale circulation (Figure 4.2).

Today, the average mixing timescale of the deep sea is more than 1000 years. So normally, exchanges with the deep sea are a slow affair. The intensity and depth of mixing are related to wind strength, which in turn depends on atmospheric temperature contrasts, temperature and salinity contrasts within the ocean, and tidal energy; the large-scale ocean circulation consequently has a considerable temperature overprint. So periods of strong temperature change are likely to be times of change in the exchange of carbon between the hydrosphere and the biosphere-atmosphere. What we see here, in simple terms, is that initial temperature changes, through ocean mixing change, may trigger carbon exchange from the deep sea into shallower layers. This brings it into communication with the atmosphere.

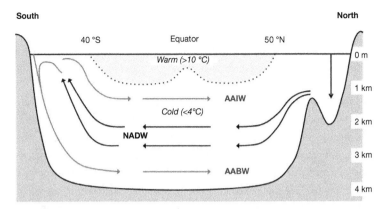

Figure 4.2. Schematic of deep-sea circulation in the Atlantic Ocean, along a section from south to north.[iii] Direct gas exchange with the deep-ocean water masses only happens at high latitudes. AAIW is Antarctic Intermediate Water. AABW is Antarctic Bottom Water. NADW is North Atlantic Deep Water.

iii. Rohling, E.J., The oceans: a deep history. Princeton University Press, 272, 2017. ISBN9781400888665

The gas-exchange between atmosphere and ocean is also affected by sea-ice cover, especially in the southern ocean. Cold periods with large areas of sea-ice cover have reduced gas-exchange between the ocean and the atmosphere. Reduction of the sea-ice cover with warming allows more gas-exchange. Once the deep-sea carbon reservoir through all these complex interactions begins to affect the atmospheric CO_2 level, a powerful positive feedback system becomes established. Initial warming triggers CO_2 release from the ocean into the atmosphere, which causes more warming, which causes more CO_2 release, and so forth, until a new equilibrium is reached between CO_2 levels in the ocean and atmosphere.[69]

The same processes operate in the reverse direction, except that it takes longer to accumulate carbon in the deep sea; once mixing is reduced, a long time is needed to let decomposition of sinking dead organic matter drive the build-up of carbon in deep waters. As a result, important CO_2 release from the deep sea may occur over a few thousand years, and storing CO_2 back into the deep sea typically takes tens of thousands of years. Note that I have presented a very schematic picture, and that I completely ignore the carbonate compensation process. The latter complicates the simple picture but in essence does not change the direction of change toward the end result—it merely makes the resultant CO_2 changes larger.

The previous paragraph raises an important question whether the rate of ocean mixing is always as slow (over thousands of years) as discussed. In other words: could we simply ignore the ocean when it comes to the fate of human emissions and climate change in the next century? The answer is a resounding no, since the ocean is more than just the deep sea. There are also CO_2 uptake and release mechanisms related to its so-called "saturation state" in shallow and intermediate levels, and these operate on much faster timescales, from years to centuries. We refer to these as CO_2 equilibration processes. When looking at century timescales and at carbon changes that are not too far away from pre-industrial levels, a useful rule of thumb applies to these faster interactions. This is that any gain or loss of carbon through the atmosphere will be divided between the atmosphere and ocean in roughly equal $(1/1)$ proportions,[2,70] due to the continuously ongoing equilibration between the upper ocean layers and the atmosphere. Detailed reconstruction in the fifth assessment report[71] of the Intergovernmental Panel for Climate Change (IPCC) suggests that the atmosphere/ocean proportions may be closer to 1.5/1.0, but we will for simplicity of argument use a 1/1 ratio. Although this changes the numbers, it does not change the argument.

The 1/1 case implies that net human carbon emissions have been roughly twice as large as what we actually measure in the increase of atmospheric CO_2 levels since the start of the industrial revolution—almost half (actually about 40%) of it was taken up by the ocean (compare Figures 1.1 and 1.2). This split is not perfect, and it will change for large CO_2 deviations as more time goes by, since this allows further interactions within the carbon cycle; in particular, exchange with the deep sea and carbonate compensation. The split also changes as ocean temperature, saturation state, and mixing timescales change far from their pre-industrial values. But those reservations aside, it's a useful first-impact estimate to be kept in mind, especially when talking about atmospheric CO_2 capture and storage in the near future (see section 6.2).

This brings us to the terrestrial biosphere, the living and dead biomass on land. While variations in the terrestrial biosphere cannot offset all of the carbon fluctuations caused by exchanges with the vast oceanic carbon reservoir, they do take care of a considerable proportion. During cold periods (in the extreme: ice ages) the size of the terrestrial biosphere shrinks—it holds less carbon. During warm, lush periods, the biosphere expands and therefore holds more carbon. So on a transition from a cold period to a warm period, expansion of the biosphere fixes a lot of carbon, removing it from the atmosphere-ocean system. In a reverse change, contraction of the biosphere releases a lot of carbon to the atmosphere-ocean system. The atmospheric reservoir is small enough that its carbon content is noticeably affected by such changes. In contrast, the oceanic reservoir is so enormous that it experiences only subtle changes in response to the biosphere changes. Over the last ice-age cycle, the terrestrial biosphere first contracted and later again expanded by about 500 GtC, although there is considerable uncertainty about the exact number (to be further discussed shortly).

More recently, deforestation by humans has made a significant impact. It has caused the land-biosphere to contract by an estimated net value of 100 GtC[2,72] because forests of trees hold much more carbon than grasses, grains, pulses, and shrubs. This has taken place over many centuries, but was strongly accelerated during the last century due to the human population explosion. At the same time, there have been other human-caused impacts involving the terrestrial biosphere that compensated for part of the impact of deforestation, such as land-use changes. The net human-caused emissions from the terrestrial biosphere have been estimated at 30 GtC in the IPCC fifth assessment report.[71] This net carbon shift (= emission) from the biosphere to the atmosphere and ocean adds to our fossil fuel emissions, giving a total of about 430 GtC by the end of 2017. As stated above, just under half of this net carbon flux (about 155 GtC[71]) has ended up in the ocean over that time, causing measurable ocean acidification.[67,70]

If we ignore human-driven processes, then the timescales for natural (non-assisted) biosphere contractions and expansions range from hundreds to thousands of years. For example, estimates of the increase in the carbon content of the terrestrial biosphere over the 10,000-year long termination of the last ice age range from about 300 to 1100 GtC.[73,74] These estimates included both living and dead material, thus also including things like peat-land development. The range implies an average rate of 0.03 to 0.11 GtC per year, although intervals of more rapid change may have alternated with periods of little to no change. Human-caused change in the terrestrial biosphere concerned a net release of 30 GtC concentrated in about 200 years, or 0.15 GtC per year, which is at the fastest extreme end of estimates for natural changes in the terrestrial biosphere.

During the natural carbon cycles of the past ice ages, over several hundreds to many thousands of years, not only the terrestrial biosphere is important, but also interactions with other, slower feedbacks, such as—especially—exchange with the deep sea and carbonate compensation. It is now well established that the "re-appearance" of carbon during the terminations of ice ages—in the form of atmospheric CO_2 increases by about 100 ppm (involving about 200 GtC) and terrestrial biosphere expansions by about 500 GtC—was dominated by carbon release from the deep sea, boosted by impacts of the carbonate compensation process. As the oceans warmed up, circulation changed, sea-ice retreated, and mixing of glacial deep water into shallower layers brought the deep waters in touch with the atmosphere for equilibration.[75] Further impacts came from release of gases (especially methane) that were trapped in the extensive permafrost areas of the ice age. In summary, many of the major, slow carbon cycle feedbacks were involved over these longer timescales, not just equilibration with the surface and intermediate layers of the ocean.

The slow feedback processes complicate projections of what may happen with CO_2 in the future. While the past CO_2 changes include many (slow) processes that are not so relevant for the next 100 years of our future, these processes will definitely become major players on a longer timescale. We should not forget this: the politically much focused-upon year 2100 is not some magical barrier by which time the climate system will stop responding to our actions. The slow responses, combined with further slow responses related to ocean warming and albedo change due to ice-sheet reduction, mean that the climate system will continue to adjust over many centuries to come, *even if emissions are stopped from today.*[76]

Another implication of the involvement of many of the major, slow carbon cycle feedback processes in past natural climate changes is that it becomes difficult or impossible to pinpoint cause-and-effect relationships just by looking at which change came first and which change came a bit later. This is because, throughout these changes, the system was caught in a tightly interlinked feedback loop with some "slack" in the connectivity between the various processes (the stretchy chain in the "feedback bicycle" discussed in section 3.1). A good example is the question of whether warming or CO_2 rise started first during the deglaciation. In reality, both respond to each other, and it doesn't really matter very much how the cycle was started.

This neatly brings us to the heart of the matter of natural climate variability: if the system developed along a certain pathway because of the action of feedback processes, then what set it off in the first place? What caused the initial perturbations of the system during natural (before humans) climate changes? Here we have to think about several different causes. The main ones are astronomical variability in the Earth-Sun configuration that changes the intensity and distribution of insolation (ISWR), events such as major volcanic eruptions, comet or asteroid hits, and variability in the intensity of solar radiation. These are discussed next.

4.2. ASTRONOMICAL VARIABILITY

Earth has gone through an endless, almost rhythmic succession of climate fluctuations, and the ice ages were part of that. These cycles are more regular, and span much shorter timescales, than the plate-tectonic cycles discussed before. They mainly took place on distinct, regularly repeating timescales, referred to as "periodicities," which span 21 thousand, 41 thousand, 100 thousand, and 400 thousand years. There are some longer overlying periodicities as well, of 1.2 million and 2.4 million years, but we will focus on the first set, as they concern the fundamental processes.

These periodicities are determined by astronomical parameters, which regulate different aspects of the configuration of Earth in its orbit around the Sun. Climate is sensitive to both the total amount and the latitudinal and seasonal distribution of solar radiation received by Earth. Three astronomical cycles are of relevance: the eccentricity cycle, the obliquity cycle, and the precession cycle. There is a beautifully accessible account of the history of discovery of the ice ages and their astronomical origin in *Ice Ages: Solving the Mystery*.[77] In short, drawing upon 100 years of research on astronomical variations and their influence on climate, the Serbian engineer Milutin Milankovitch determined the fluctuations in the intensity and distribution of solar radiation onto Earth.[78]

One of Milankovitch's special contributions was the calculation of past insolation variations over various discrete latitude bands. In the scientific community concerned with climates of the past, the three main astronomical cycles are often referred to as the "Milankovitch cycles." Ever since his work, research has continued to improve and update the astronomical calculations. We will go through the three relevant astronomical cycles, and then sum up their key impacts.

The cycles of eccentricity and precession need to be discussed together, as their influences are closely related. Before these cycles can be discussed, however, let's have a refresher about what determines the seasonal cycle (Figure 4.3). One year marks the time needed for Earth to

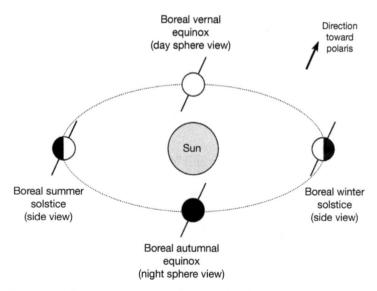

Figure 4.3. Schematic presentation of a seasonal cycle. Note the importance of the fixed direction in space of the rotation axis on these short time scales (today toward Polaris). If the axis were not tilted relative to the plane of orbit, then there would be no seasons.

complete one rotation around the Sun. On short timescales of years to centuries, the position of Earth's rotational axis is more-or-less fixed in space relative to the plane of the planet's orbit around the Sun, with the North Pole today pointing toward the star Polaris. Consequently, there is a season in which one of the poles is tipped away from the Sun (the winter hemisphere), while the other pole is facing the Sun (the summer hemisphere). Six months later, this situation is reversed.

Let's follow one rotation, from the northern hemispheric (also known as boreal) perspective. We start at the northern winter solstice, the shortest day on the northern hemisphere, when the North Pole is turned most away from the Sun. The northern winter solstice marks the start of winter on the northern hemisphere. The next notable point along Earth's orbit is the northern spring or vernal equinox, the start of spring on the northern hemisphere. During an equinox, the Sun reaches its zenith directly over the equator, and the boundary between the illuminated and dark half-globes passes through both poles. Half a year after the winter solstice, the earth reaches the northern summer solstice, the longest day on the northern hemisphere, when the North Pole lists most toward the Sun—this marks the start of northern summer. Next, the northern autumnal equinox is reached, marking the start of northern autumn.

The eccentricity cycle concerns variations in the shape of Earth's orbit around the Sun, from near circular to elliptical or oval shaped. An ellipse has two focal points, and as the ellipse transforms to a circle, the two focal points approach one another. In a perfect circle, the two focal points overlap and together define the center-point of the circle. The Sun occupies one of the focal points of the Earth's orbit; the other is empty. During an eccentricity maximum, when the orbit is notably elliptical, Earth in one of its yearly revolutions around the Sun therefore passes a point nearest the Sun, known as perihelion, and a point furthest away from the Sun, or aphelion (Figure 4.4). When the orbit is near circular—an eccentricity minimum—Earth's distance to the Sun is virtually constant through the year. The eccentricity of Earth's orbit changes in a cyclic manner, with approximate periods of 100,000 and 400,000 years.

The precession cycle is related to the fact that Earth's rotational axis relative to the plane of its orbit around the Sun is not fixed in space when viewed over long timescales of thousands of years, but wobbles like the

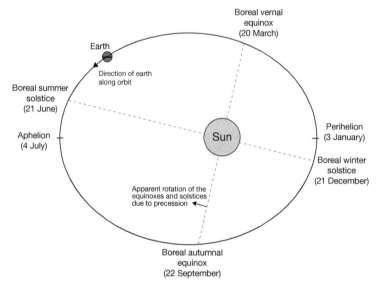

Figure 4.4. Schematic "top view" of earth's orbit around the Sun (not to scale). Also indicated are the current dates at which the earth reaches the solstices and equinoxes, and the dates at which it reaches aphelion and perihelion. The direction is given of the shift along the orbit of the solstices and equinoxes, caused by precession. (The seasons on the southern (austral) hemisphere are exactly half a year out of phase, since the South Pole points in the exact opposite direction to the North Pole. Hence, the northern winter solstice is the southern summer solstice, the northern vernal equinox is the southern autumnal equinox, the northern summer solstice is the southern winter solstice, and the northern autumnal equinox is the southern vernal equinox.)

axis of a spinning top (Figure 4.5). Here, we are not talking about changes in the angle of the axis relative to the plane of orbit—which is discussed under the third astronomical cycle, obliquity—but about changes in the direction of the axis in space. Essentially, the precession cycle causes the North Pole, which today points toward Polaris, to point toward Vega (which then becomes the North Star) after half a precession cycle, and back toward Polaris again after a complete precession cycle (Figure 4.6). A full cycle of precession takes 26,000 years. However, other complications in

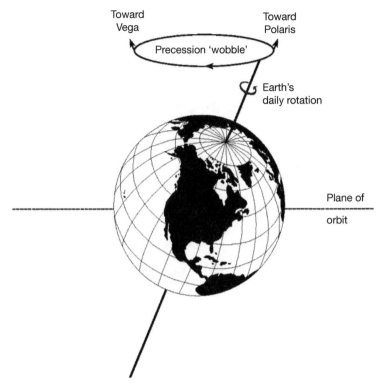

Figure 4.5. Earth's precession wobble. One revolution takes 26,000 years.

the Earth-Sun motions come into play; the entire Earth orbit itself slowly rotates around the Sun, about once for every four precession periods. As a result, the precession cycle expresses itself in the insolation onto Earth in two dominant periodicities: a major one of about 23,000 years and a minor one of 19,000 years. As a first approximation, we often use an average periodicity of 22,000 years.

The cycle of precession influences climate by causing the dates of the solstices and equinoxes to slowly shift along the orbit. A quarter of a cycle ago, or about 5,500 years ago, perihelion occurred near the northern autumnal equinox. Half a cycle ago, or about 11,000 years ago, perihelion occurred close to the northern summer solstice. Three quarters of a cycle ago (about 16,500 years ago), perihelion coincided with the northern

Precession Maximum

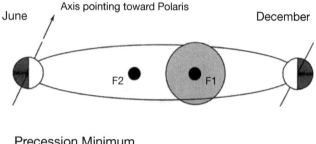

Precession Minimum

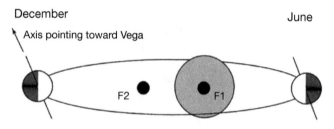

F1: Focal point one of the elliptical orbit (sun)

F2: Focal point two of the elliptical orbit (empty)

Figure 4.6. Summary schematic to demonstrate the differences between precession maxima (as today) and precession minima. Note the exaggerated eccentricity of Earth's orbit with two focal points, of which one is occupied by the Sun. The direction in space of the Earth's axis has changed from pointing toward Polaris in the precession maximum, to pointing toward Vega in the precession minimum.

vernal equinox. And a full cycle ago, the situation concerning precession was similar to the present.

The climatic impacts of precession and eccentricity need to be viewed together. In its slightly elliptical orbit, Earth today reaches perihelion (closest to the Sun) at around the northern winter solstice—these two events are near January 3 and December 21, respectively. Earth reaches aphelion (furthest from the Sun) at around the northern summer solstice—July 4 and June 21, respectively (Figure 4.4). When the orbit approaches a circle, the distance difference between perihelion and aphelion is negligible, but with some eccentricity, the solar radiation on illuminated places of the globe will be somewhat more intense in northern winter (southern summer) than in northern summer (southern winter). This weakens the northern hemisphere's seasonal contrast, and strengthens that on the southern hemisphere.

As we saw before, progress through the precession cycle shifts the distribution of the seasons around the elliptical orbit. Half a precession cycle ago, the situation was reversed relative to that observed today. At that time, perihelion almost coincided with the northern summer solstice and aphelion with the northern winter solstice. In that configuration, the seasonal contrast on the northern hemisphere is increased, and that on the southern hemisphere is weakened. In short, the precession cycle governs the seasonal insolation contrast, but its impact on climate depends on the degree of eccentricity of the orbit. In a circular orbit, precession has no impact. In times of maximum eccentricity, precession achieves maximum impact.

The third astronomical cycle is that of obliquity, or tilt, of Earth's rotational axis. More specifically, it concerns the gradually changing angle of Earth's rotational axis relative to the perpendicular to the plane of Earth's orbit (Figure 4.7). This angle changes from 22.5 to 24.5 degrees and back again, over a period of about 41,000 years. Today, the angle is about 23.5 degrees. As a result, the Sun stands directly overhead at about 23.5° north latitude during the northern summer solstice. This is the maximum north latitude where the Sun at any one time in the year reaches a directly overhead position, and it is called the Tropic of Cancer. During the northern winter solstice (southern summer solstice), this condition is reached at about 23.5° south latitude; the Tropic of Capricorn.

On a perfectly spherical Earth, the obliquity cycle would therefore shift the position of the Tropics between 22.5 and 24.5° latitude. But Earth is not perfectly spherical, and the actual values are nearer 22.04 and 24.45°. In addition, obliquity affects the amount of sunlight received at high polar latitudes. For strong tilt, the poles receive more sunlight, and the Sun's rays also reach the polar surface at a less shallow angle, which decreases

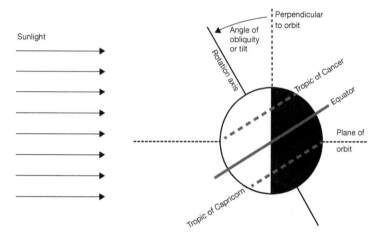

Figure 4.7. The relationship between obliquity, or tilt, and the positions of the Tropics of Cancer and Capricorn. The tilt of Earth's axis relative to the perpendicular to the orbital plane varies on a cycle of 41,100 years between about 22.5 and 24.5 degrees. Today, the tilt is about 23.5 degrees. For clarity, the angle is exaggerated in this diagram.

reflection and so increases the surface absorption of heat. In consequence, obliquity has particularly strong impacts on polar climates.

The three main astronomical cycles have limited net impact on the global average amount of ISWR (incoming short-wave radiation) that reaches Earth. From that viewpoint, the astronomical variations might almost be ignored. But that would be a mistake, because the astronomical parameters do play a major role in regulating the distribution of ISWR between hemispheres and between low and high latitudes. As discussed, precession and eccentricity are very important for seasonal changes, and obliquity is very important for high versus low latitude gradients. On a yearly averaged basis, the high versus low latitude gradients in terms of absorbed ISWR fluctuate over a range of about 7 W/m² even without any feedback responses. For scale, note that, during the last 500,000 years, albedo and greenhouse gas feedback responses amplified this yearly averaged gradient from astronomical variability to a total gradient of about

33 W/m² (Ref. 51). The importance of astronomical variations clearly lies in the fact that they drive initial climate change through seasonal and spatial absorbed ISWR gradients. The climate system's own internal feedback mechanisms then cause strong amplification of the initial disturbances into the observed total global responses.

4.3. LARGE (SUPER-)VOLCANIC ERUPTIONS AND ASTEROID IMPACTS

Along with ash, which settles quite quickly, large volcanic eruptions eject mostly water vapor, CO_2, and sulphur-dioxide (SO_2) gas into the atmosphere. The water vapor input becomes part of the hydrological cycle and condenses out swiftly. As discussed before, water vapor in the atmosphere acts as a key fast-acting feedback to temperature. CO_2 emissions from volcanoes worldwide give an external carbon input that is often estimated at about 0.1–0.15 GtC per year, although some estimates are a bit higher at about 0.5 GtC per year.[79] Over geological history, external carbon input from volcanic processes has been very important for climate through accumulation over millions of years or more, or over tens to hundreds of thousands of years during exceptionally violent events. But volcanoes do not make much of an impression relative to human-caused emissions today. At about 10 GtC per year, the current human-caused emissions are at least 20 times, and possibly up to 100 times, greater than the carbon emissions in the form of volcanic CO_2.

Volcanic emissions of SO_2 comprise an estimated 13 million tons of sulphur, or 0.013 GtS, per year. Although this is a relatively small proportion in the atmospheric sulphur cycle,[80] the episodic and regionally concentrated nature of this input still is important for global climate. Large, explosive eruptions emit SO_2 in a regionally concentrated manner into the stratosphere, thus creating aerosols of minute droplets of mostly sulphuric acid. These can remain in the stratosphere for a couple of years and cause both ISWR reflection and OLWR absorption. In addition, the sulphur compounds are destructive to stratospheric ozone. The net effects on climate are complex. Generally there is a pattern of global cooling, such as the cooling of about 0.5°C for the year after the eruption of Mount Pinatubo in 1991, but regionally there can be winter warming triggered by temperature changes in the lower stratosphere.[81] Importantly, these effects are short-lived, with timescales of a few years.

Occasionally, really big and explosive (super-)volcanic eruptions occur, such as the Toba eruption in Indonesia about 75,000 years ago, the Campanian eruption in Italy about 40,000 years ago, the Oruanui eruption of Taupo Volcano in New Zealand 26,500 years ago, and the Santorini eruption in Greece about 3,600 years ago. These may have affected climate over periods of up to several years, and much of their impact was due to amplifying feedback processes within the climate system. A more recent major volcanic eruption, the Tambora eruption of 1815, led to historically documented climate consequences, notably a "year without summer."[82] Overall, such major volcanic events have been distributed in a more-or-less random manner through time,[83] producing climate events that punctuate the regularly repeating astronomical climate cycles.

To produce a consistent change, long-lasting clusters of volcanic eruptions would be needed in time. There is no evidence that such organization of really major explosive eruptions has occurred over sustained periods of time. This does not mean that, over geological time, there have been no times of major volcanic activity. Instead, such episodes did exist, lasting hundreds of thousands to millions of years. Good examples are the emplacements of so-called Large Igneous Provinces (LIPs),[84] such as the Deccan Trap basalts of about 66 million years ago[52]. Commonly, those events did not concern volcanism of the explosive kind, which is much more effective in producing stratospheric aerosols. During such periods, therefore, the impacts of long-term external carbon emissions are thought to have dominated.

Finally, there is little doubt that asteroids or comets impacting on Earth can have devastating climatic consequences. The best-known example is the impact of a Mount-Everest-sized asteroid that terminated the reign of the dinosaurs 66 million years ago. Such big asteroid hits can set off major climate changes, notably through a so-called "impact winter" scenario.[85] But they are an essentially random, and fortunately rare occurrence. As a result, I will not consider this type of event any further.

4.4. VARIABILITY IN THE INTENSITY OF SOLAR RADIATION

Here, we encounter a complication. On the one hand, a lot is known from precise measurements of recent variations in Total Solar Irradiance, or TSI, which is the total intensity of output from the Sun. TSI is what I referred to before as total solar radiation, and which amounts to about 1360 watts per square meter (W/m^2). TSI has been measured by satellites above the Earth's atmosphere for the last 40 years or so, which revealed that TSI fluctuated over a total range of less than 2 W/m^2 over that period.[86,87] A 2 W/m^2 range in TSI translates into a global mean ISWR average variability of 0.5 W/m^2, before reflection. That is only 0.15% variability in ISWR. But let's not dismiss solar variability too easily, and first consider the "on the other hand" argument.

TSI has been found to vary along with the number of visible sunspots; the more sunspots are visible, the higher the TSI. Sunspot counting has been done in earnest since about the year 1610, and sunspot area recording since 1874.[86,87] The TSI estimates from such methods, and from longer extensions over the last 11,500 years based on measured variations in radiocarbon and beryllium-10 production rates, reveal typical cycles of variability of about 11 years, as well as some longer cycles (Figure 4.8). Intriguingly, many records of past climate change show the same cycles and patterns.[22,88,89] This has led to suggestions that solar output variations somehow manage to cause climate variability.

This raises a critical question: how large were the solar variations? During the Little Ice Age, roughly in the years 1400 to 1850, there were three periods of zero, or very low, sunspot activity: the so-called Grand Minima. During the most recent of these, the Maunder Minimum of about 1645–1715, average TSI seems to have been between (most likely) 0.8 and (at most) 3 W/m^2 lower than the average today if we account for all uncertainties.[87] Therefore, in the deep Maunder Minimum low of solar output, ISWR was only 0.06 to 0.22% weaker than today. Records show that the Maunder Minimum is among the very deepest lows in TSI, and modern values among the highest highs, within the last 10 thousand years (Figure 4.8).[87]

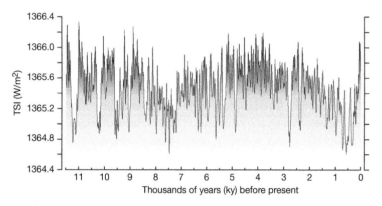

Figure 4.8. Total Solar Irradiance (TSI) fluctuations over the last 12 thousand years.[iv,v]

iv. Viera, L.E.A., Solanski, S.K., Krivova, N.A., and Usoskin, I. Evolution of the solar irradiance during the Holocene. Astronomy and Astrophysics, 531, A6, 2011. doi: 10.1051/0004-6361/201015843.

v. Solanski, S.K., Krivova, N.A., and Haigh, J.D., Solar irradiance variability and climate. Annual Review of Astronomy and Astrophysics, 51, 311–351, 2013.

The question of how such minor changes in ISWR could have expressed themselves in the climate system to the extent that they clearly seem to show up in records of past climate change is a very profound one.[90] There still is no complete answer. Yet the answer will be very important to everybody—from the scientist who tries to understand exactly which processes account for what response to the person who wishes to brush climate change aside on the basis that, "It's all about natural variability, not about human disturbance."

If we think about it a little, then we find a real nugget of information. Yes, TSI influences may have been small, most likely about 0.15 W/m^2 and at most 0.5 W/m^2 change in ISWR (after reflection). But the very fact that climate responses in a wealth of reconstructions of natural, pre-industrial change seem to convincingly coincide with such minor solar output variations—with the Little Ice Age as a critical, historically documented example—means that small changes in the radiative balance of climate can somehow trigger noticeable changes in climate through amplification by feedback mechanisms within the climate system.

This nugget places an interesting limitation on anybody's argument (and I have heard many) that, "Surely the observed changes are simply due to the Sun; the Little Ice Age was the result of a solar minimum, and the modern warming is nothing but the result of recovery from that solar minimum." The limitation is that, by this reasoning, the person implicitly and—you bet—unwittingly accepts that very small unbalances in Earth's energy balance have major consequences for climate, since the solar changes that they are referring to are known to be small.

In other words, the person infers a very high sensitivity of climate to change in the energy balance. Ironically, climate researchers, who are often rebuked by the same people as alarmists who overstate the sensitivity, actually infer much lower sensitivity. This is simply because the researchers ascribe only a fraction of the temperature change since the Little Ice Age to TSI changes. In addition, they consider a broader range of changes to the energy balance, from volcanic influences and from human-caused greenhouse gas and aerosol effects. In one fairly typical exercise like that, researchers attributed only 0.1°C of the 1°C temperature rise since the Little Ice Age to TSI.[87] Here we have a clear case where the scientific assessment is actually much less dramatic than the implication of a throw-away statement that is often used to try and make researchers look alarmist. In fact, it turns out that the alarmist hat is squarely on the person who wishes to explain all change in terms of just solar variability.

It is also important to take a brief moment to think about a matter of scale. When we talked above about 0.15 W/m^2, to a maximum of 0.5 W/m^2, of energy unbalance from ISWR (after reflection), it may have sounded almost insignificant because (after reflection) ISWR amounts to about 240 W/m^2. But a better comparison is with the total energy balance shifts associated with recent ice-age cycles that involved about 5 to 7°C global temperature change. These shifts amounted to only about 10 W/m^2. So 0.015 to 0.05 times 5 to 7°C would give a reasonable first guess for the TSI influences on global average temperature. This yields a likely value of 0.08°C, to a maximum value of 0.35°C. Those are not entirely insignificant numbers, and they actually agree rather well with values from more sophisticated model assessments.[87]

4.5. RECAP AND OUTLOOK

In this chapter about the causes of past natural climate changes, we started with CO_2 variations owing to the operation of Earth's carbon cycle. We considered two main types of change: one involving fluxes of external carbon, and one involving only redistribution of carbon within the active climate system's carbon-cycle components (mainly atmosphere, oceans, and biosphere). We saw that even the most dramatic natural injections of external carbon into the climate system during the last 100 million years or more were considerably slower than that caused by humanity today, and that it took nature hundreds of thousands of years to clean up those emissions again. We also found that natural slow feedback processes within the climate system will cause any fast CO_2 rise to cause warming that grows and persists for at least several centuries, even if emissions were stopped completely.

In terms of actual drivers of natural climate changes, external carbon injections were discussed, followed by changes in seasonal and spatial ISWR gradients due to astronomical variations, which lead to an initial response that then is strongly amplified by feedback processes within the climate system. We argued that volcanic eruptions influence climate mainly through sulphate aerosols, while long-lasting periods of exceptional volcanicity are known to have caused notable greenhouse gas forcing at specific times in deep geological history. We only briefly discussed large asteroid impacts with evident implications for climate.

Finally, we spent some time considering total solar irradiance variations. We found these variations to be weak, but surprisingly often represented in records of climate fluctuations. The link between weak forcing and measurable impacts is not yet understood, but we did consider the implications of making too big a deal of it. Attributing such climate shifts as the Little Ice Age completely to solar variations would imply an unrealistically high climate sensitivity to radiative forcing changes.

Up to this point, we have concentrated on understanding what drives climate change from a perspective of natural variability. We found that natural climate variability was mostly driven by initial perturbations that were relatively slow and of relatively modest sizes and rates. These modest perturbations then caused significant climate variations because of the action of feedback processes, which act over almost all timescales, from nearly immediately to millions of years (Figure 3.2). Now it is time to see what's changed since the start of the industrial revolution.

[5]

CHANGES DURING THE INDUSTRIAL AGE

Most of the 1°C temperature change since the start of the industrial revolution has occurred in the last six decades (Figure 1.1). The warming is evident in all independently monitored timeseries of global temperature.[4] The general warming trend has been overprinted by variability on a lot of different timescales, largely because of internal (re-) distributions of heat within the atmosphere-ocean system. The world ocean, with an average depth of 3700 m, has more than 1000 times the heat capacity of the atmosphere.[91] Even just the upper 700 m that are in effective exchange with the atmosphere have 200 times the heat capacity of the atmosphere. As a result, even a tiny fraction of a degree centigrade change in just the upper ocean represents an enormous amount of heat. This means two things: first, atmospheric temperature can be substantially affected by almost undetectable changes in the ocean; and second, ocean heat gain calculation requires very precise temperature measurements. Precise measurement series for the ocean only exist since about 1960. Let's have a look at what atmospheric and oceanic heat gains tell us about the Earth's energy balance since the industrial revolution.

The roughly 1°C rise of Earth's surface temperature during the industrial age, with more than two-thirds of it since about 1960, represents the "realized" response to forcing. Using standard values for global climate sensitivity to radiative forcing, we can determine that this 1°C warming corresponds to a component of climate forcing of roughly 1.1 to 1.3 W/m².

In contrast, the ocean is such a vast reservoir to heat up that it has not yet realized its full warming—ocean warming will therefore continue to develop over many decades to centuries even if we managed to "freeze" all radiative forcing agents at their current levels. Since 1960, the heat content of the upper 2000 m of the ocean has increased by roughly 27×10^{22} joules in about 55 years.[92] This is an enormous number; namely 27 followed by 22 zeroes. For comparison, the most powerful nuclear detonation ever had a yield of about 22×10^{16} joules. So, ocean warming since 1960 equates to an energy uptake of roughly 1.2 *million* times that explosion, or 60 of such explosions *every day* since 1960. We can do a ballpark calculation for the radiative forcing that would result in such heat gain.

One watt is one joule per second, so 27×10^{22} joules over 55 years equals about 156 trillion watts, where one trillion is one million millions. The world ocean's surface area is 361 trillion square meters[93] and the average surface area for the layer between 0 and 2000 m depth is roughly 90% of that,[94] or 325 trillion square meters. Therefore, we find a rough estimate that the time-averaged oceanic heat gain in 55 years following 1960 was equal to an average forcing of some 0.5 W/m^2. Sophisticated recent assessments find similar values (0.4 to 0.6 W/m^2) for the upper 2000 m,[95] or 0.53 to 0.75 W/m^2 based on the upper 300 m,[96] while most recent whole-ocean estimates for 1991 to 2016 suggested increasing values between 0.72 and 0.94 W/m^2.[97]

From these estimates, let's just say that the representative ocean value is roughly 0.5 W/m^2. Overall, we then find that Earth has been responding to a total radiative forcing of 1.1 to 1.3 W/m^2 for which surface warming has already been realized, plus about 0.5 W/m^2 for which surface warming is not yet fully realized because it is being used to slowly warm up the ocean. This totals to 1.6 to 1.8 W/m^2 (Box 5.1).

Our crudely estimated range of 1.6 to 1.8 W/m^2 is in pretty good agreement with estimates of the net radiative forcing of climate between 1.13 and 3.33 W/m^2 by the Intergovernmental Panel on Climate Change (IPCC),[98] which are made by adding the radiative impacts of all known processes (Figure 5.1). So the numbers add up; the amount of radiative climate forcing that can be estimated from measurements of atmospheric and oceanic heat gain is in good agreement with the amount of forcing that one would reconstruct from addition of all the known contributing

Box 5.1.

A global average of roughly 1.6 W per square meter may not sound like much, but it's an awful lot of heat. Earth surface spans about 510 million square kilometers, or 510 trillion square meters. Let's try to get a feel for the amount of heat represented by 1.6 W/m² over such a surface area. One large gas patio heater produces 12,000 W when going full blast, so one such patio heater on an area of 7500 square meters gives an average of 1.6 W/m². That means that 1.6 W/m² as an average over the entire world surface would be equal to 68 billion large gas patio heaters going full blast, day and night, on Earth (in other words, about 10 large heaters going full blast, day and night, for every single person on Earth).

processes of radiative change. In view of this, it is futile to go on denying that warming has taken place, or stating that we do not understand the critical drivers behind it.

However, if still more persuasion is needed, then consider the relationship between temperature change and the total amount of carbon emitted (Figure 5.2). That relationship turns out to be very straightforward: it is virtually linear, and very tightly defined. This, incidentally, makes it the most intuitive relationship for making temperature projections with respect to future emission amounts.[99] Evidently, we don't need to capture all the finest details to make the big lines clear to all who are willing to look without bias. It would seem more productive to accept what's measured, and to then devote one's energy to working out what can be done about it.

As mentioned before, the IPCC has considered the drivers of radiative change in detail (see Figure 5.1[98]). It goes too far to go over all that again, as the report is publicly available. It is clear that CO_2 and CH_4 are the critical ones to consider for positive change in the radiative forcing (and aerosols, and surface albedo due to land-use changes, for negative effects). Since the onset of the industrial revolution, CO_2 increase has added about 2 W/m²,

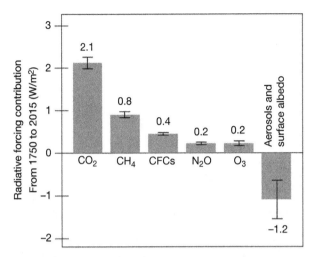

Figure 5.1. Simplified overview of the major contributions to changes in the radiative forcing of climate from the start of the industrial revolution to the present.[i] For a more complete analysis, see ref. 2. Note that aerosols and surface albedo had a negative contribution, because increased atmospheric aerosol loads and clearing of less reflective woodlands in favor of more reflective agricultural land have made the planet more reflective since the industrial revolution. The double T-shaped bars I ndicate uncertainty intervals, and the written values give the mean estimate for each contribution.

i. Hansen, J., Sato, M., Kharecha, P., von Schukmann, K., Beerling, D.J., Cao, J., Marcott, S., Masson-Delmotte, V., Prather, M.J., Rohling, E.J., Shakun, J., Smith, P., Lacis, A., Russell, G., and Ruedy, R., Young people's burden: requirement of negative CO$_2$ emissions. Earth System Dynamics, 8, 577–616, 2017.

ii. Figure SPM-05 in http://www.ipcc.ch/report/graphics/index.php?t=Assessment%20 Reports&r=AR5%20-%20WG1&f=SPM

and CH$_4$ has added about 1 W/m^2 (Figure 5.1). Looking at just the last 15 years, CO$_2$ accounts for more than 80% of the added greenhouse gas forcing, and—if we remain highly dependent on fossil fuels—CO$_2$ will be the dominant driver of future global temperature change.[2,72] This is why we only briefly discuss CH$_4$, before passing swiftly on to a more detailed discussion of CO$_2$.

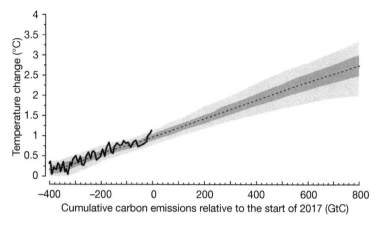

Figure 5.2. Temperature change as a function of cumulative (total sum) carbon emissions. Values are shown relative to the start of 2017 and include model-based projections into the future.[iii] The heavy line shows the observations, and the dashed line shows the median through the data and future projections (the median is the level with 50% of values below it, and 50% above it). The dark gray shading indicates the 68% confidence interval, and the light gray shading indicates the 95% confidence interval.

iii. Goodwin, P.A., Katavouta, A., Roussenov, V.M., Foster, G.L., Rohling, E.J., and Williams, R.G., Pathways to 1.5 °C and 2 °C warming based on observational and geological constraints. Nature Geoscience, 11, 102–107, 2018.

Methane (CH_4) concentrations in the atmosphere have nearly tripled since pre-industrial times. This increase includes both external carbon input and carbon redistribution within the hydrosphere-biosphere-atmosphere system. Methane is readily oxidized into CO_2 and water vapor, and removal of methane mainly depends on this chemical break-down, rather than on uptake of carbon into other reservoirs, as is the case for CO_2. About one-third of methane emissions into the atmosphere has come from natural sources, from wetlands, termites, the oceans, and per-mafrost reduction. Two-thirds have come from entirely human sources, mainly fossil fuels and intensive livestock farming. In addition, there have been considerable additions from waste treatment and landfills, biomass burning, rice agriculture, and biofuels.[100,101]

The CO_2 increase almost exclusively concerns addition of external carbon. Overall, atmospheric levels have increased from about 280 ppm to above 400 ppm since the onset of the industrial revolution (Figures 1.1, 2.1).[102,103] Over the ice-age cycles of the last 800 thousand years, measurements of CO_2 concentrations in fossil air bubbles trapped in the ice sheets of Greenland and Antarctica show that CO_2 levels fluctuated between 180 and about 280 ppm (Figure 2.1).[103,104] So, during the last 800,000 years, they never reached as high as they are today. Also, the CO_2 increase since the industrial revolution (120 ppm) is larger already than the natural CO_2 swings associated with the ice ages (100 ppm). CO_2 levels before 800 thousand years ago can only be measured indirectly, but still suffice to indicate that CO_2 levels close to the present 400 ppm last occurred on Earth during the Pliocene Epoch, at around 3 million years ago,[60,105,106] when the planet was some 2°C warmer than today.

The modern CO_2 rise of about 120 ppm has taken place over about 150 years. That is an average of about 0.8 ppm per year. But the CO_2 rise is speeding up, and in the last few years the levels have increased by some 3 ppm per year. In contrast, the ice-core data reveal that natural CO_2 levels changed at typical rates of 1 ppm *per century*, with maxima of about 3 ppm *per century*. It appears that the CO_2 increase in recent years is about 100 times faster than the fastest natural CO_2 increases of the preceding 800 thousand years. Similarly, we saw before that the very highest estimates for the rate of rise during some of the most catastrophic natural events of external carbon injection in the geological record (the CIEs) are about 3000 ppm in about 4000 years, or 0.75 ppm per year. That is at best similar to the average rate for the past 150 years, and some 4 times slower than the rate of CO_2 rise in the last few years. These comparisons strongly suggest that the modern CO_2 rise is not natural—even during its most catastrophic events, nature did not achieve the rates of CO_2 rise that we see today.

I am often asked how we can be so sure that humans have caused the industrial-age CO_2 rise, as opposed to natural processes. There are several ways by which this can be demonstrated. First, and easiest, we can demonstrate that about 95% of the external carbon input that underpins the modern CO_2 increase is caused by emissions from the burning of fossil fuels by looking at economical records of fossil fuel use (Figure 1.2).[2] The other 5% resulted from cement manufacturing. In doing these assessments, it was found that our carbon emissions have been roughly twice as much as needed to explain the atmospheric CO_2 rise.[107] Therefore, there is no doubt that our emissions easily suffice to explain the rise. As mentioned before, most of the "missing" half of the emitted carbon has disappeared into the oceans, where it causes acidification.

Second, a convincing indication that most of the modern CO_2 increase is caused by fossil-fuel emissions comes from a widespread change toward lower atmospheric carbon isotope values, as measured in a variety of materials.[108,109,110] Due to processes involved in the accumulation of carbon in plants and subsequently in petroleum maturation, fossil fuels have low carbon isotope values. The burning of fossil fuels releases this signal, which then causes the measured shift. Note that records from the ocean show a similar trend to increasingly lower carbon isotope values.[110,111] Given that the trend is seen everywhere, it has to be caused by a global change in the carbon isotope values due to external carbon injection, rather than a redistribution of isotopes between reservoirs.

Third, fossil fuels are millions of years old and therefore have no radiocarbon activity, since radiocarbon decays away in less than 100,000 years. As a result, drops over the last 200 years in radiocarbon measurements made on growth rings of trees[112] are a clear signal of old (fossil) carbon injection into the atmosphere due to emissions.[113]

In summary, data from different and independent lines of study prove beyond reasonable doubt that (a) CO_2 levels during the industrial age have risen to levels not seen in 3 million years; (b) CO_2 levels during the industrial age have risen at extremely high rates, recently some 100 times faster than any rate of rise known from the last 800 thousand years, and several times faster than during the most catastrophic natural events of external carbon injection; and (c) these industrial-age CO_2 changes were caused to a very large proportion (about 95%) by fossil fuel burning.

The rest of this chapter evaluates some critical aspects of the consequences of the excessive industrial-age CO_2 rise.

5.1. DIRECT EFFECTS

First and foremost among the direct effects of the changes in Earth's energy balance since the industrial revolution is surface warming, both on land and in the ocean. Warming on land is particularly noticeable. It has caused clear poleward expansion of climate zones in both hemispheres, with widespread consequences. For example, poleward expansion of the subtropical climate zones has caused increasing aridity at the poleward margins of these zones.[2,114] The tropical regions have also noticeably expanded,[115,116] and there has been a distinct contraction of polar conditions in the Arctic region, which we will revisit later in this section.

The shifting of climate zones has profound impacts. These consequences are not only related to the latitudinal shifts of climate zones, but also to widespread upward displacement of climate zones toward higher altitudes in mountainous regions, which compresses the ecological niches of high-altitude adapted flora and fauna.[117] For example, some 75% of marine species have experienced about 1000 km poleward shifts in their habitats.[118,119] More than 50% of land-based species have experienced range shifts of as much as 600 km in a poleward direction, and some 400 m in an upward direction.[120] Overall, it has been established that human influence has—for many reasons, of which climate change is only one—pushed species extinction rates to modern levels that are 1000 or more times faster than natural rates of species extinction in the past.[52,121]

Even more devastating direct impacts of warming await us in a few decades if we continue along our business-as-usual pathway of emissions (that is, if we continue as we have in the last decade or so). Humans, as well as most other large mammals, are susceptible to stress and eventually death from heat. Heat stress develops at around 38°C at 0% humidity and near 20°C at 100% humidity, while heat becomes lethal at around 50°C at 0% humidity and near 30°C at 100% humidity, especially if exposed for about 20 days or more without relief.[122] Many tropical and subtropical regions of the world are perilously close to these limits already. Under a business-as-usual scenario, the lethal temperature threshold will be exceeded virtually *every day of the year* (300+ days per year) throughout large swathes of Amazonia, tropical West Africa, the southern half of India, and Indonesia by the year 2100.[122] We are only a few decades away from times in which—under a business-as-usual scenario—heat stress and heat death conditions will reach and exceed the threshold of 20 consecutive days.

Other large mammals will be affected similar to humans. As a result, we should expect major impacts on food stocks as well as human populations, especially because retreating into cooled/air-conditioned environments is not an option for cattle, pigs, and other animals. In addition, wild ecosystems will be affected, and so will agriculture and marine systems, the effects of which are likely to lead globally to falling yields and productivity.

In the oceans, warming is severely affecting coral reefs already.[123] Early in the 1980s, coral bleaching events happened once every three decades or so. Today, it is happening once every six years or so, or more frequently in some places. Bleaching occurs because the corals are stressed by high water temperatures. This causes the corals to expel their symbionts, which are algae that photosynthesize by trapping energy from sunlight using pigments. Ejection of the algae therefore causes the coral to lose its color and become transparent. This in turn allows the bright, white calcium-carbonate skeleton to be seen through the coral polyp, so that it appears bright white, or "bleached." When bleaching events are infrequent, the corals can take up new symbionts and regain their colorful appearance. But recovery typically takes a couple of years, and may be incomplete— some sections of reef may die. When bleaching events follow each other too rapidly, there is not enough time for recovery and most of the reef will die. Scientific reef monitoring has shown that we've arrived at this critical point in many reefs around the world.[124] It's a problem that cannot be emphasized enough, given that coral reefs are home to about a third of all biodiversity in the oceans, while the oceans cover more than two thirds of Earth's surface area.

Warming has some important further consequences: an increase in evaporation, and a higher water-vapor transport capacity in a warmer atmosphere. This has been driving a notable increase in the occurrence of regional droughts where evaporation dominates, and flooding where precipitation dominates. In addition, the increased latent heat transport in the atmosphere (that is, the heat used in evaporation and released during condensation), and the intensified gradients between warm and cool regions, underlie generally more energetic conditions with increases in the intensity of storms. Recent work suggests an intriguing response in which weaker storms get weakened, and stronger storms intensified.[125]

Hurricanes are a case in point. Note that hurricanes are known as typhoons in the western Pacific, but that I here refer to all as hurricanes. Development of hurricanes requires a sea surface layer warmer than 26.5°C, and limited vertical shear in the winds over that sea surface;[126] that is, wind speeds need to remain roughly the same at sea level and higher altitudes. Under the right conditions, hurricanes develop from tropical storms that gain intensity. As air is sucked faster and faster into the storm's intensifying low-pressure cell, the rotation of the Earth eventually deflects the airflow into a powerful circulation around a central eye: the infamous hurricane winds. Hurricane rainfall and winds are extremely destructive by themselves, and the winds in addition drive abrupt and devastating changes in local sea levels. These so-called storm surges can reach many meters,[127] cause extensive coastal damage, and can drive massive inland flooding when they penetrate up rivers and breach levees.

The poleward expansion of climate zones in both hemispheres that we encountered before is matched by poleward expansion of warm ocean regions, which has happened by about 60 km per decade over the past 30 years.[128] This increase in the area affected by hurricanes at least partly results from the greenhouse gas increase,[129] and brings a threat of hurricanes to regions where they never or rarely occurred before. Some modeling work has suggested, for example, that western Europe may be increasingly caught in the crosshairs, even under moderate greenhouse gas emission scenarios.[130] To some extent, one might view that study as a "prediction" of hurricane Ophelia, which managed to reach the British Isles in 2017.

Over the past century, the oceans have been warming, and this trend has accelerated in the past three decades.[131] It is expected to cause a change to fewer but more intense tropical cyclones and hurricanes.[132] It also increases the rainfall from all tropical cyclone and hurricane events.

We may therefore expect a triple whammy in flooding potential from increased storm surges, enhanced total rainfall amounts, and expansion of hurricane-sensitive regions.

Warming has also caused a distinct contraction of polar conditions in the Arctic region, associated with reductions in the regions affected by sea ice cover over sea and permafrost on land, and with resultant contraction of tundra regions. This has triggered a further climate feedback, since melt-back of permafrost and warming of polar oceans has caused release of methane (CH_4) that used to be locked below and inside the frozen sediments for thousands of years.[133,134,135] Methane is a powerful greenhouse gas; molecule for molecule, the greenhouse potential of CH_4 is over 25 times stronger than that of CO_2. But it is relatively short-lived in Earth's oxygenated surface environments. Methane's average lifespan is only a few years, as it readily oxidizes into CO_2. CO_2 is less potent, but is long-lived in the atmosphere (around a century), with about one fifth lingering for many thousands of years.

Some research has proposed a relationship between reduction of Arctic sea ice and snow cover and changes in the pathway of the subpolar jet stream. Along with general climate-zone shifts, this may underlie increased occurrences of extreme weather events, such as sharp wintry spells over central and eastern North America and central and northwestern Europe, spells of anomalously warm winter conditions over Alaska, and European/Eurasian heat waves.[136,137] Without substantial reduction of greenhouse gas emissions, it is estimated that extreme weather could kill as many as 152,000 people per year in Europe alone by 2100, compared with 2700 deaths per year in recent times.[138] The overwhelming majority of those deaths would be heat-related.

Although it doesn't concern the radiative forcing of climate, it is important to mention one final direct consequence of our carbon emissions. It concerns the increasing amount of CO_2 taken up by the ocean, which drives ocean acidification.[52,67,68] Together with warming, this has a detrimental effects on a wide range of organisms, especially those with carbonate skeletal parts, such as oysters, mussels, clams, corals, and microplankton with carbonate skeletons.[139,140,141] It will require timely development of locally relevant measures to limit the commercial impacts, for example on shell fisheries.[142]

Geological studies of past carbon emission events (the CIEs) reveal that ocean acidification at those times impacted upon, but did not terminate such organisms, possibly because the changes were slow enough to allow organisms to adapt to the increasing ocean acidity.[143] As we have seen before, human carbon emissions and the associated ocean acidification take place at faster rates than those of even the most dramatic CIEs. It is feared that this rapidity may prevent a similar adaptive response in many marine species.

For more impacts, and more detail on those briefly discussed above, I refer the reader to the plethora of reports that have addressed the various direct responses and their potential for increasing climate extremes, including anomalous heat waves, precipitation events, extreme storms, storm-track relocations, etc. See, for example, reference 2, and of course the extensive and authoritative IPCC reports. In the following, I will instead focus on larger-scale responses that develop over longer timescales, and for which a context can be formulated from geological studies of natural climate variations, before human interference. These can be organized under two headings: climate sensitivity and sea-level change. We will consider them in turn.

5.2. GLOBAL RESPONSES AND CLIMATE SENSITIVITY

For any increase in radiative forcing, the climate system must respond. As we have seen, the resultant warming causes an increase in OLWR that eventually brings the planet back into energy balance.

Warming will affect both the ocean, and land, but their warming responses have different timescales. Warming of the ocean is a slow process; only the surface layers of the ocean absorb incoming radiation, which warms them up, and only the surface of the ocean exchanges heat with the atmosphere, both in a direct sense and through so-called latent heat exchange that involves evaporation and condensation. As the winds gradually mix the upper ocean layers down to depths of about 700 m, the surface effects drive changes in those upper 700 m or so within several decades to a century or two. Heat exchange with deeper oceanic layers, however, requires mixing and large-scale circulation, and proceeds very slowly; the deep-sea circulates with a typical timescale of a thousand years. So ocean warming starts to affect the upper several hundreds of meters within decades, which is rapid for the ocean but still quite slow on human timescales, while the deep ocean will only adjust over many centuries to more than a thousand years.

Warming of the land is a different matter; it is a very rapid process. Upon brief reflection, almost everybody will know this from personal experience, as sea temperatures change relatively little between seasons, while land temperatures virtually immediately follow even day-night changes. This is why, at similar latitudes, maritime climates such as that over western Europe are characterized by much smaller fluctuations over time than continental climates such as that over Canada.

In short, land temperatures adjust almost directly to changes in the radiative forcing of climate, while ocean temperatures take many decades to many centuries to adjust. This has important implications for considerations of climate change at any moment in time. Before these implications can be evaluated, however, we first need to consider the further consequences of any initial warming. These are driven by activation of the complex feedback processes within the hydrosphere-biosphere-atmosphere system that were discussed before (Figure 3.2).

Positive feedbacks amplify the response. Negative feedbacks dampen the response. Some of the feedbacks act rapidly, in a matter of years or less (Figure 3.2). Examples are the water-vapor and cloud feedbacks, snow and sea-ice feedbacks, and (natural) aerosol feedbacks. Other feedbacks are slow, acting over many decades to centuries, through to many thousands of years (Figure 3.2). For example, moderately swift responses may be expected from vegetation changes (several decades to century-scale). These influence both the land-surface albedo, with considerable reflectivity changes for bare sand and rock *versus* grassland *versus* woodland, and for dust transport into the atmosphere from the blowing out of desert sands, a natural aerosol component. Surface albedo feedback from land-ice changes (here we mean ice that survives through the year, not seasonal snow and ice) has response timescales from decades to centuries for mountain glaciers, to thousands of years for major continental ice sheets. Parts of the carbon-cycle feedback, most notably carbon storage and release from intermediate and deep ocean water masses, including carbonate compensation, operate on timescales from centuries to many thousands of years.

So while ocean warming may be a slow process on human timescales, it definitely is not the slowest response in the climate system. Any disturbance to the radiative budget will set in motion a chain of events that reverberates over many centuries to thousands of years.[76] This puts the global warming targets for the year 2100 from the COP21 Paris climate conference of 2015 in a new light: the forcing (greenhouse gas emissions) allowed for such warming will also trigger slow responses, and these will then cause climate to keep warming for centuries to come after 2100, until eventually the entire climate system has equilibrated to the forcing change. The end result after a century or two when the surface ocean has finally warmed up is about 1.5 times the warming that the politically agreed targets "allow" for 2100, even if no further changes were made to the radiative budget. If we allow the slow feedbacks to adjust too, meaning that we wait a good few centuries, then the eventual temperature will be about double the warming that the targets "allow" for 2100, even if no further changes were made to the radiative budget. We'll get back to these delayed responses a bit later again, as they are a crucial aspect of total climate change in response to any disturbance. The climate system's response does not have a cut-off date at 2100, as one might think from the political discussions. Instead, it keeps changing as the fast responses get followed up by the moderately fast, slow, and very slow feedbacks (Figure 3.2).[76]

It is now important to view things in terms of the resultant temperature change per W/m^2 of radiative forcing of climate, which is known as climate sensitivity. Although climate sensitivity is a response to whatever radiative forcing, it is often reported in terms of the temperature response per forcing amount equivalent to a doubling of CO_2. For this, each doubling of the CO_2 levels represents almost $4\,W/m^2$ of forcing. The IPCC[71] reports climate sensitivity (let's call it CS) as the temperature response to a doubling of atmospheric CO_2 concentrations after action of the fast feedbacks, but I here prefer to work with values in degrees C per

W/m^2 of radiative forcing because I think it's a more intuitively clear way of discussing climate sensitivity.

Either way, there is a complication. As you will gather from the fact that feedback processes operate on virtually all timescales (Figure 3.2), we have to make a decision about which timescale we are discussing. As mentioned, CS is the value after action of the fast feedbacks; that is, mainly the water-vapor, cloud, snow, and sea-ice feedbacks.[59] In essence, therefore, CS gives the mean temperature change that will be achieved on decadal timescales. But the climate system will not reach a global average temperature that is properly, or even nearly, in equilibrium with the radiative forcing that fast because ocean temperature changes take much longer.

While the ocean is still in the process of warming up, the ocean is effectively still too cool at any time. This in turn keeps average global temperature values down. Once the ocean has finally warmed up to a reasonable depth—normally we consider this to be several hundreds of meters—average global temperatures will therefore end up higher. The climate sensitivity after this ocean temperature adjustment, but not yet including the slow feedback influences (Figure 3.2), is known as "equilibrium climate sensitivity" (ECS). As we have seen from the discussion before, ECS will be roughly 1.5 times larger than CS. Where CS gives temperature change that will be achieved in decades, ECS gives the eventual temperature change after a century or two.[76]

Incidentally, a further term can be encountered in the literature. It is the so-called Earth System sensitivity (ESS), which describes the total temperature response relative to forcing by CO_2 change alone, and commonly views things over many centuries to thousands of years, so that most of the important slow feedbacks have contributed as well. We will not expand on this topic any further here.

ECS is estimated by the IPCC at a likely range of 0.4 to 1.2°C for every W/m^2 of radiative climate forcing.[71] This estimate is mostly based on climate-model assessments, with some additional information from past periods such as the Last Glacial Maximum (about 20,000 years ago). In separate detailed assessments of ECS estimates from a broad suite of geological observations, it was found that these consistently fell within a likely range of 0.6 to 1.3 °C per W/m^2 throughout the past 65 million years.[59,60] The strong agreement between these independent evaluations brings out an important message: the basic range of ECS is quite well understood. Current efforts are focused on trying to reduce the range of uncertainty. Several different methods are being investigated to narrow the range of the ECS estimates, both using climate modelling[144,145] and using model-based assessment of combined geological and historical data.[99]

The 2°C "limit" of warming that was internationally agreed at the COP21 Paris climate conference in 2015 can also be assessed from the perspective of climate sensitivity. For a correct appreciation of this 2°C "safe limit of global warming to avoid dangerous consequences," the distinction between ECS and CS is absolutely essential. The 2°C value has taken on a dominant political life of its own, although it has very limited scientific grounds[146,147]. Early work inferred that avoidance of 2°C warming would allow roughly one doubling of CO_2 concentrations from pre-industrial levels (to about 550 ppm), but it was soon found that it instead requires stabilization of CO_2 levels below 400 ppm.[148] Current CO_2 levels are at that limit already, and continue to increase by some 3 ppm per year (Figure 1.1).

In the following, we first briefly consider the total amount of carbon emissions still allowed before we hit the 2°C warming level, and we also consider the Paris agreement's more aspirational goal of limiting warming to 1.5°C. Next, we consider these limits in terms of CS and ECS. And we discuss the importance of timescales of change.

The easiest way to get a sense of the total amount of emissions still allowed (relative to the beginning of 2017), before we hit 1.5 or 2°C warming relative to pre-industrial temperatures, is using the simple relationship—with well-expressed uncertainties—in Figure 5.2. This graph shows results from one study, but other studies using very different approaches give similar results. It reveals that the 1.5°C value will be reached after another 195 to 205 GtC have been emitted, and the 2°C value after another 395 to 455 GtC have been emitted. We know from Figure 1.2 that current business-as-usual emissions are about 10 GtC per year, so that the 1.5°C and 2°C warming thresholds are set to be reached within only two and four decades, respectively.[99] This is much sooner than "allowed" under the Paris climate agreement. So we must find solutions to reduce our emissions, and even to remove carbon from the climate system. But before we can go any further into those topics, we need to do a bit of deeper thinking about setting specific warming targets for the year 2100. What do such targets actually mean?

As we have seen, focusing on 2100 as if all climate change processes would simply stop there overlooks especially the slow process of ocean warming. If the 2°C limit is taken to apply to the year 2100 (essentially, the CS response)—as it too frequently is when targets are discussed on the political scene—then ongoing slow-response-driven change toward a new equilibrium (the ECS response) implies that the temperature change over the longer term will reach 3 to 4°C, even if absolutely no further radiative forcing is applied.[2,76] In consequence, the correct interpretation of a "2°C safe limit to global warming" would be one that considers this as the limit to the total (long-term) response. It then follows that the truly acceptable value for 2100 is only 1 to 1.3°C.[2] Given that we have already gone through 1°C warming (Figure 1.1), this is an extremely pressing problem: we have much less wiggle-room than we might have thought.

It is instructive to place these arguments within the context of our recent warm period—the Holocene. This is the period that started about 11,500 years ago, following the end of the last ice age. Some work infers that about 0.25°C of warming since pre-industrial times brought us back to the warmest it had ever been during the Holocene.[2] This is confirmed as far as mean values were concerned, but if we also account for uncertainties, then it appears that modern temperatures are only just beginning to exceed the past temperature peak to a significant level.[149] This means that industrial-age warming of about 1°C has now brought the planet to a point where it is definitely becoming warmer than before during the Holocene. However, this 1°C of industrial-age warming predominantly represents only the CS response, and cannot be taken as the endpoint. It will continue to increase in the next century or so, to 1.5°C or more, even without any more radiative forcing, as ocean temperatures are going through their slow adjustment in completing the ECS response. And it will grow even more, to 2°C or so, as the slow feedbacks play out over subsequent centuries.

Clearly, we cannot rest on our laurels because of what I sometimes hear: "Modern warming is not that dissimilar from the past Holocene peak." The Holocene peak came about because insolation was a little bit higher than today and had developed very gradually over thousands of years. In contrast, today's warming is caused by a rather large anomaly in the Earth's energy balance that results from a faster-than-ever rise in greenhouse gases. In fact, this rise has been so fast that the climate response is completely out of equilibrium, effectively trying to play "catch-up" as fast as it can. From the above discussion of CS versus ECS, we now know that this "catching up" means that warming will not stop even if the forcing of climate does so. To make matters worse, it seems highly unlikely that the climate forcing itself will be stabilized in the near future. Instead, it will keep on increasing as emissions continue. Thus, real warming will eventually exceed even the ECS estimates listed above. In consequence, the statement above about comparison with the Holocene peak is a fallacy. Curbing CO_2 to an absolute maximum of 400 ppm is imperative to avoid more than 2°C total warming.

And then the argument gets even more difficult: some researchers (including me) consider that a 2°C warming limit is too generous already, and that aiming for 2°C even on the longer term will have "dangerous" consequences[2,150]. Even a target of 1.5°C may be too generous. A broad suite of geological data of past natural climate changes indicates that we would better limit warming to 1°C,[2] which implies CO_2 levels of at most 350 ppm—that is, more than 50 ppm lower than today. This assessment was not an isolated one, or extreme: it is only slightly more pessimistic than other assessments.[148,150] For a 2°C target, most remaining fossil fuels need to remain untouched[151]. For a 1°C target, the same applies, and in addition there are CO_2 reduction requirements.[2]

One further aspect requires attention. Up to now, we have considered warming responses to radiative forcing of climate in a globally averaged sense only. But in reality, hardly any place on Earth would be affected by a global average value, unless it is by sheer coincidence. Instead, the climate response at any specific location will be dominated by regional influences. These make it stronger than average in some places, and weaker than average in other places. In a way, this is the same as how hardly anybody will be exactly as tall as a given group's average height (except by coincidence); most people in the group will be either shorter or taller than the average. The regional contrasts exist because different climate forcing and feedback processes have different influences across the world; for example between low and high latitudes, or between land and ocean surfaces, or between highlands and low-lying regions.

A key influence that causes regional climate responses to differ from the average is so-called polar amplification. This refers to a much stronger-than-average temperature sensitivity to changes in the global radiative forcing at the poles than at lower latitudes. It is dominated by surface albedo (reflectivity) changes that result from expansion and contraction of the snow and ice cover in those high latitudes. In cooling climates, snow and ice cover expands, making the high latitudes more reflective to ISWR, which in turn causes more regional cooling. Conversely, snow and ice cover contracts in warming climates; this reduces the reflectivity at high latitudes, which in turn causes more regional warming.

On centennial timescales, polar amplification makes temperature sensitivity at high latitudes some 50 to 150% stronger than the global average temperature sensitivity (Box 5.2). In other words, for a given radiative forcing of climate, temperature at the poles may be expected to change 1.5 to 2.5 times more than the global average value that we get from climate sensitivity. If such sizeable regions change more than the average,

Box 5.2.

Note that the commonly reported polar amplification values are those for the current century, or one to two centuries further into the future. So it is the amplification factor that is relevant to CS or ECS. These values do not include the impacts of changes in continental ice sheets, as those are part of the slow feedbacks (over thousands of years) that are not counted within CS or ECS. Instead, the slow feedback influences are accounted for separately, and make polar amplification even stronger when considering the very long timescales of ESS. In practice, this means that, when we estimate polar amplification from geological records, we need to make a correction for the land-ice effects to find estimates that are relevant on timescales of CS and ECS.[59,60]

then other large regions must be changing less than the global average. The regions that show lower temperature sensitivity than the average value are the low latitudes (notably between about 30° North and South latitudes).[51] In those regions, temperature changes will be more subdued than the global average value that we get from climate sensitivity.

5.3. SEA-LEVEL CHANGE

Global sea-level changes on historical to ice-age timescales are dominated by changes in ocean temperature, and by transfer into the ocean of water that was previously locked away as ice on land. On longer timescales of many millions of years, the volumetric capacity of ocean basins fluctuates due to changes in ocean spreading rates[52], but those processes are so slow and gradual that they are not considered here.

Temperature is important because ocean water expands when it warms up, causing sea level to rise; the mass of water in the ocean remains the same, but it requires more volume because it expands. This influence amounts to between 0.2 and 0.6 m per °C of top-to-bottom ocean warming, and it is immediate with ocean warming.[71]

Ice-sheet reduction transfers extra mass (ice and water) from land into the global ocean. Over ice-age cycles, the complete response of this influence resulted in roughly 20 to 25 m per °C of global warming (compare sea-level change with global temperature change in Figure 2.1). This component of sea-level rise relies on retreat of continental ice sheets, which happens over many centuries to millennia.

In view of the above, it is clear that the immediate sea-level response since the industrial revolution has been dominated by the rapid thermal expansion component, along with some additional fast melt-water input from mountain glaciers. But, even in the long-term future, the impact of these components will remain limited to less than a meter because of limited warming of the oceans and the limited size of mountain glaciers. By comparison, the longer-term response related to mass loss of the great continental ice sheets has been slow to get going because it takes considerable time to change the rate of ice-sheet processes (like trying to set a heavily laden freight-train into motion). Once moving, however, ice-sheet processes have great momentum. Thus, they will over time drive sea-level rise to many meters above the present, even if the forcing of climate were reduced again, like a heavy freight train on full speed taking a long time to stop even when the brakes are applied. It is important to consider these longer-term consequences in more detail.

As polar regions warm up in response to the previously discussed global average warming with superimposed polar amplification, the large continental ice sheets at high latitudes become increasingly affected by climate change. These ice sheets are today found over Greenland, and over West and East Antarctica. If they were to melt completely, then both Greenland and West Antarctica would add enough water to the global ocean to cause roughly 7 and 5 m of sea-level rise, respectively. The East Antarctic ice sheet is much larger, and has the potential to add roughly 53 m of sea-level rise.

As anybody who has looked at time-lapse photography of glaciers will know, ice is not rigid. Instead, it flows very slowly. Even in large continental ice sheets, we recognize so-called ice-streams, where ice is flowing faster, like a (very) slow river. The net movement of ice streams is down toward the coast, and then out over the coastal ocean. When in water, about 90% of ice lies below the surface and about 10% above it. So when an ice stream reaches the ocean with a thickness of, let's say, 300 m, then it will keep sitting on the ocean floor as long as that is less than 270 m deep—we call this grounded ice. When the ocean floor falls away deeper than about 90% of the ice thickness, then the ice floats in the ocean, and we speak of a floating ice shelf. Ice shelves of different ice streams can combine into one larger ice shelf. The point where the ice shelf changes from grounded to floating is called the grounding line (Figure 5.3).

Ice shelves are considered to be critical regions of weakness in the behavior of ice sheets. In particular, they are very important for the processes that determine how fast ice streams can transport ice toward the ocean, since grounded ice shelves provide a lot of friction and thus buttress their feeding ice streams—in plain words: they put the brakes on their feeding ice streams. The most critical issue is the sensitivity of the grounding line to melting of the ice from underneath, which in turn is dominated by ocean temperature. When polar oceans warm

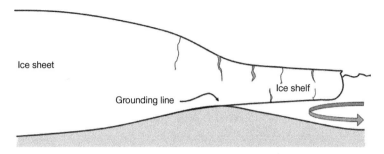

Figure 5.3. Schematic of an ice shelf. Also illustrated are the grounding line, and under-ice ocean circulation (which drives under-ice melting).

up, under-ice melting intensifies and the grounding line retreats landward.[152,153,154] This leads to weakening and potential break-up of the ice shelves. When ice-shelf collapse happens, the feeding ice streams speed up because the brakes have been taken off, and this increases the amount of ice that moves from the continental ice sheet into the ocean. In turn, this causes sea-level rise.

Once the ice sheet has retreated completely away from contact with the ocean, the processes of mass transport from land to sea are expected to slow down, because the warmer ocean can no longer directly weaken the ice, so that the ice is affected by much slower land-based melting. Air temperature increases around ice margins also drive increased ice-mass loss, especially when combined with ocean temperature increases.[153] And processes such as hydrofracturing from the surface, and ice-cliff collapse at the ice front, are increasingly being looked at to explain events with the highest rates of ice-mass loss.[155,156]

By now, it will be evident that some of the most vulnerable ice sheets are those with large sectors that are grounded below sea level. The West Antarctic Ice Sheet (often abbreviated to WAIS) is almost entirely grounded below sea level. To make matters worse, the Southern Ocean has been warming at an alarming rate during the last couple of decades.[157] Warming has caused large ice shelves around WAIS to suffer significant thinning and grounding-line retreat, and several have disintegrated.[152] This is why the scientific community is most concerned about WAIS with respect to potential future sea-level change. As mentioned before, WAIS holds enough water for about 5 m of sea-level rise, and most of it is vulnerable to rapid collapse.

The Greenland Ice Sheet (GrIS) also has several important sectors that are potentially affected by ice-shelf reduction and ice-stream acceleration, but different slopes of the shallow sea-bed (relative to that around WAIS) may limit these influences.[158] On this basis, perhaps roughly 1 m of sea-level rise could come from GrIS rapidly. Thereafter, it might be expected to slow down and be affected by slow land-based melting and calving only. But there is another process at work in Greenland, which makes the ice sheet much more vulnerable to climate warming than thought before, and which involves interaction of melt water with the weak sediments discovered at the base of the ice sheet.[159] So GrIS could potentially provide more sea-level rise than roughly estimated above.

The East Antarctic Ice Sheet (EAIS) is very thick and most of it is very stable. In some sectors, however, it has ice-shelves that are grounded below sea level, buttressing major ice streams that drain considerable portions of EAIS. The largest of these sectors holds enough ice for a 19 m sea-level rise, and there are other sectors as well.[160] These are the more vulnerable sectors of EAIS, where rapid loss could occur when ocean waters warm up.[161] The ice volumes in these sectors may add up to more than 20 m of sea-level rise.

Adding up all the relatively "fast" ice-sheet contributions discussed above, we get up to some 5 m from WAIS, 1 m or more from GrIS, and 20 m or more from EAIS. This means roughly 25 m of sea-level rise through fast processes. That range is familiar to most researchers of past climate, as it is within the range of estimates for the amount by which sea level stood higher than today during the warm Pliocene, and specifically during the interval of 3 to 3.5 million years ago, the most recent period during which CO_2 levels were close to the modern level of 400 ppm.[42,75,105,106] It is also the approximate value up to which the modern-type (relatively fast) sea-level sensitivity to CO_2 change applies, and above which the sea-level sensitivity to CO_2 changes is reduced (because then, the slow ice processes dominate).[105]

The Pliocene studies, within their longer-term geological background, show us two things. First, sea level may be expected to rise by many meters in a climate with CO_2 levels sustained at 400 ppm or more. Given several hundreds to thousands years of time, there is little doubt that this will happen. Second, the Pliocene work shows that our present climate state falls squarely within the range of relatively fast sea-level responses, and that it will remain so until sea level has risen at least 9 m, and according to most reconstructions between 12 and 32 m, above the present level.[42,105]

But how fast is fast? This is an important question because processes that are fast in a geological context are often still excruciatingly slow in human terms. So we need to clarify whether we might expect all of the fast sea-level response to occur in, say, one century, or whether ice flow is so slow that even a so-called rapid collapse would still be slow in human terms. In other words, what do we know about the timescales involved in the sea-level response?

Computer-based numerical models, based on the most recent under-standing of ice physics, will be essential for making detailed predictions for the future. However, the objective here is not to try and make accurate predictions, but instead to gain a realistic first sense of the scale and speed of the potential sea-level responses to warming. A good first step is to look at past, natural (pre-industrial) periods of warming and sea-level rise, to obtain estimates of the natural timescales and rates of sea-level rise in re-sponse to warming.

Use of geological observations means that we get a range of estimates from real-world examples of how fast things may change under totally nat-ural forcing. Because nature has so neatly played these examples out for us in the past, we can have no doubts that it possesses the mechanisms to make similar changes in the future. But, as we have seen before, the current (anthropogenic) forcing of climate is extraordinarily fast. We can use com-parison between our range of sea-level rise estimates from the natural past and modern observations to see whether the measured modern sea-level changes are also extraordinarily fast. In other words, we can see whether our geological (natural) perspective still applies for the future, or whether we have entered uncharted territory without natural precedents.

A few limitations apply to the use of geological data. One is that geo-logical data for sea-level change cannot distinguish between the causes of change; thermal expansion and mass-addition are rolled into one number. This is not much of a problem because thermal expansion is negligible rel-ative to mass-addition when considering big sea-level changes over more than a few decades to centuries or millennia. Another limitation is that geological data come with rather large uncertainties. But honest assessments of these uncertainties allow us to formulate confidence intervals to the answers. Where we report so-called likely ranges (commonly denoted with ± values), we refer to 68% confidence limits around the central estimates. These limits delineate the range within which the actual value of an uncertain estimate will fall with 68% confidence.

Geological data for the past 500 thousand years show that maximum natural rates of sea-level rise were dependent on the amount of ice that existed on the planet before each rise (see note 10 of Figure 2.1). It's rather logical: the more ice there is, the more can be lost to the ocean at any given time. To compare past data with current sea-level trends, we must therefore focus on past events during times with approximately the same amount of ice on the planet as exists today (therefore, starting with sea-level close to the present). Doing so reveals maximum natural rates of sea-level rise of about 1 ± 0.5 m per century, and duration estimates of about two to seven centuries for the build-up of each event from its onset to its peak rate of rise (see note 10 of Figure 2.1).[43] Used in a straightforward mathematical relationship, such values can roughly illustrate what we might expect of sea-level rise in the industrial age, if it still was entirely governed by the natural processes that also governed the natural (before humans) sea-level rises.[162]

Such work suggests that measured sea-level rise during the industrial age[163,164] may seem modest, but still was rapid enough to track the upper limit of the 68% confidence zone of natural rises. In other words, the current rate of sea-level rise is not completely without precedent in geological terms, but it certainly is developing along a trajectory that is super fast already, relative to natural changes. As ever, everything is relative, depending on how we look at it: sea level only appears to be rising slowly to us humans because ice sheets are slow to respond to change, and we have—until now—experienced only the early stages of this ice-sheet response. The "freight-train of continental ice-sheet mass loss" has very slowly rolled into motion over the past century or so. But now that it is moving, it will only continue to gain momentum. Recent years have given clear warnings of such movement, in the form of major ice-shelf collapses, ice-stream accelerations, and intensifications of net mass-loss from ice sheets.[152, 153, 154, 165] In addition, observations have revealed strong warming of polar oceans, a likely trigger for much of the ice-sheet responses.

If we use the upper 68% confidence limit of natural sea-level rises—which the current sea-level observations seem to follow—to project into the future, then a global average sea-level of about +1 m above the pre-industrial level (which was 30 cm lower than today) is suggested for the year 2100,[162] an estimate that was supported later by more elaborate statistical analyses.[166] The expected levels then grow to just under +3 m for 2200, and +5 m by 2300,[162] and some ice-modelling studies suggest that +15 m may be reached by 2500.[156] Even if the radiative forcing of climate due to carbon emissions were stopped completely, so that we would stabilize CO_2 levels at around today's value close to 400 ppm, there would be no reason to assume that the freight-train that has finally come into motion would stop any time soon again. As discussed before, geological data indicate that for CO_2 levels maintained at 400 ppm, the slow climate feedbacks—which include the ice-sheet adjustments—will progressively adjust over many centuries, until finally a new Earth Energy Balance is achieved. By that time, conditions on Earth will be rather similar to those of the warm Pliocene, when CO_2 levels also were about 400 ppm. This includes a global temperature of 2 to 3°C higher than today, and a sea level that stands many meters above the present level. Pliocene values suggest that continued rise to even more than +20 m may be on the cards, which would take about a millennium.

If we want to put the brakes on the freight-train of continental ice-sheet mass loss, then just stabilization of CO_2 levels at 400 ppm clearly is not enough. That would merely allow the freight-train to continue over time toward a new Pliocene-like equilibrium. Instead, we would need to reduce CO_2 levels, likely to 350 ppm or lower.[2] If we do this soon enough, then we might just have a chance to slow the ice-loss processes down before they gather too much momentum. If we don't, then all bets are off. In that case, all we can do is continue to monitor sea-level rise, and use those observations to keep improving projections of its eventual endpoint, so that we can use those projections to develop coastal adaptation and flooding-impact-mitigation plans.

Unfortunately, it is unlikely that emissions will be stopped from today. In consequence, CO_2 levels will not even stop anywhere close to 400 ppm.

On the contrary, CO_2 levels are rising faster than ever, so that the radiative forcing of climate keeps increasing. If future sea-level observations were to show rates of rise shifting beyond the geological confidence envelope (that is, toward geologically unprecedented rates), then that would strongly suggest a shift of the system into uncharted territory. Such unprecedented high rates of rise might for example arise from a WAIS collapse. And as we have seen, some parts of East Antarctica are now also considered to be more vulnerable than previously thought. This is why ice-sheet modelling is currently placing much emphasis on improving grounding-line physics and other processes in ice-shelf break-up, to better understand potential ice-sheet collapse.[156]

5.4. COMMON REACTIONS TO THE GEOLOGICAL PERSPECTIVE

Many people that I talk to about sea-level change experience one or both of the following reactions.

The first reaction is that the given sea-level rise numbers sound un-believably large. This reaction is understandable from the viewpoint of each person's personal experience. After all, the complex, large-scale, and long-timescale processes involved are rather abstract and alien to people who are not used to thinking about change on a planetary scale. Here it is important to remember that the numbers are not fictional: they simply arise from carefully collected and verified geological evidence, and stand even when allowing for uncertainties. This evidence-base clearly indicates the following:

1. The Pliocene gives a useful approximation of a climate state that is fully adjusted to the radiative forcing from CO_2 at 400 ppm.
2. The speed of sea-level rise today is still (just) within the range observed in geological data. Therefore, we do not (yet) need to consider any processes outside those involved in the geological changes.
3. The speed of modern sea-level change is toward the highest values seen in geological data. My personal opinion is that

today's fast increase in climate forcing may have optimized the rates of response, and that faster responses will require unprecedented processes, such as a collapse of the WAIS. But other researchers disagree, and instead argue that rates of response will continue to increase with increasing warming.

4. The deceptively slow and small sea-level rise of the past 150 years is not proof of a deficit in our understanding. On the contrary, it confirms our understanding that development of mass-loss from ice sheets requires a long time and that—once initiated—the ice-sheet contributions to sea-level rise will continue to dominate for many centuries to come (the freight-train argument).

The second reaction is that, at a century or more, the discussed timescales are all so long that they hardly seem relevant to the decision-making process for our future. The feeling is that *Surely, engineering solutions will be found within such timescales.* As a Dutchman, from a nation where battle with the sea is almost in our genes, I like to agree with this, but only in principle and with two caveats.

The first caveat is that I disagree that centennial timescales would be irrelevant. Instead, they define the long-term background change, on top of which shorter-term issues such as storm surges and tidal effects need to be considered. The longer timescale may seem irrelevant for decision making in a context of development-project lifespans, but perhaps project lifespans are not the best measure. Just consider that projects (buildings and other structures) in cities might be virtually completely renewed on a multi-decadal timescale, but that these renewed cities generally remain in fundamentally the same location as the initial settlements. The initial settlements and their supporting agricultural and transport traditions date back many centuries or millennia for many European and Asian cities, or still several centuries for the more recent developments in Africa, the Americas, and Australia. Their locations were selected on the basis of strategic relevance and accessibility, and long-term sea-level rise was not taken into account. Instead of project lifespans, we should worry about such long-term commitments. And for the majority of cities, ports, industry, major coastal defenses, and surrounding infrastructure and traditions, such commitments are measured on a (multi-)centennial timescale. It is definitely relevant to look several centuries ahead.

The second caveat is that modern humanity has fixed itself in place by major infrastructure and rigidly enforced borders. As a result, regional pressures can swiftly lead to major financial, political, and even military consequences. For example, a city may need to be relocated further inland when it can no longer cope with the personal and financial risks of its changing conditions (Box 5.3).[167] How fast might relocation be done, measured from the decision stage, through planning, to actual development and resource relocation? As another example, some nations may simply run out of space and/or resources, resulting in waves of "climate refugees" who flee either from adverse conditions themselves or from conflict triggered by scarcity of resources. Addressing such large-scale logistical, financial and—most importantly—humanitarian issues cannot be left to politicized knee-jerk reactions within an election cycle of just a few years. Instead, it will require a process of thought, planning, and action that is informed by a longer-term perspective. For this process, it is especially important to understand that, once we allow substantial sea-level rise to get under way, there is *no way* in which we may even hope to stop or reverse it within the next couple of centuries.

Box 5.3.

Even for only a 0.5 m sea-level rise (currently expected at around the year 2070), the 20 global port cities with the greatest rates of population growth are all going to see increases of 4 to 13 times in population exposure to extreme sea-level events (all in developing regions: Asia and Africa). Some cities will even see a 21 times increase in population exposure. In addition, 17 out of the top 20 cities for asset exposure to extreme sea-level events will see increases of 21 to 120 times in asset exposure (all in Asia and Africa).[167]

Finally, let's briefly consider another statement I sometimes hear, namely that, since human ancestry evolved in the Pliocene, similar climate conditions evidently "cannot be bad for us." This argument is simply not applicable to the actual problem that is being addressed. During the Pliocene, small numbers of nature-adapted hominins evolved over an extended period of time and in a very specific set of environments with relatively slowly changing conditions. Today, we are concerned with the potential outlook for billions of people all over the world who are closely tied into fixed infrastructure, in a system characterized by climate changes that are unravelling at an exceptionally fast pace. It is a non-argument.

5.5. RECAP AND OUTLOOK

In this chapter, we have considered changes since the onset of the industrial revolution. We saw that atmospheric CO_2 concentrations have soared to more than 400 ppm, and that global mean surface temperature has risen equally quickly by about 1°C or slightly more. We then discussed that this observed warming represents the already "realized" component of response to the change in climate forcing, and is representative of 1.1 to 1.3 W/m^2 in forcing change. We found that the much slower process of ocean warming is playing "catch-up," and that ocean heat uptake amounts to roughly 0.5 W/m^2. Thus, the total compares well with separate process-based evaluations of the radiative changes through the industrial era. This implies that we have a decent understanding of what's happening and why. We also saw why we are confident that these changes are caused by human actions.

We next discussed the major implications of warming both on land and in the sea, both in the future, and today already. This was followed by a discussion of the impacts on tropical storms and hurricanes, leading to a great increase in coastal flooding potential. Then we launched into a somewhat technical, but essential, discussion of how climate feedback processes will affect the warming response to the observed radiative changes over different timescales. This highlights a critical point: climate change won't stop in 2100, even if we stop all emissions. Slow feedbacks will continue to adjust to the radiative changes and climate responses that have happened already, and they will cause further warming after 2100. In consequence, the total temperature response will keep growing for centuries and possibly even longer. This has implications for the COP21 warming targets of 1.5 or 2°C by 2100, which we found to be too generous to avoid dangerous consequences.

Finally, we considered implications for sea-level change. We saw that, for current emissions, sea-level rise to about 1 m above the pre-industrial level may be expected by 2100 (that is, a rise to 0.7 m above the present-day level), and that further rise to +3 m by 2200 and +5 m by 2300 may be reasonably expected, with some recent studies even reporting much higher values. These major amounts of sea-level rise are related to the relatively slow dynamics of the great continental ice sheets. Once activated, these dynamics cannot be stopped. We therefore argued that CO_2 levels need to be urgently reduced to about 350 ppm to avoid activating the ice-sheet dynamics more than we already have.

So it appears that we have a problem; a big one. The prudent, responsible response would be to rapidly reduce emissions and gear up for carbon removal from the ocean-atmosphere system. But what *should* the way forward look like? Would it be sufficient to rely on Mother Nature to don her apron and clean up the mess of our excessive party? Or should we take responsibility and at least help with the cleaning? The next chapter looks at these options.

[6]

MOTHER NATURE TO
THE RESCUE?

Now we come to the key issue. Many discussions about climate change turn to the well-known fact that (very) large CO_2 fluctuations have happened in the geological past. This is then taken to imply that "we shouldn't worry: nature has seen this all before, and will somehow clean up our external carbon emissions."

The veracity of this sentiment can be tested by considering the main mechanisms available in nature for extracting carbon from the atmosphere-ocean system. These are weathering, reforestation, and carbon burial in soils and sediments. In the next section, we look at the potential of these processes. Thereafter, we consider the case for human intervention, and potential ways forward.

6.1. WEATHERING, REFORESTATION, AND CARBON BURIAL

A first mechanism by which nature has dealt with past high-CO_2 episodes is chemical weathering of rocks[52,168]. In warmer and more humid climates, chemical weathering rates are increased, and this extracts CO_2 from the atmosphere. However, CO_2 removal through weathering at natural rates is an extremely slow process, which operates over hundreds of thousands to millions of years. Given time, there is no doubt that natural weathering will be capable of eventually removing the excess CO_2, but this process is so slow that it offers no solace for the future, unless we are prepared to wait many hundreds of thousands of years. There may be some future in artificially increasing the weathering processes to remove anthropogenic carbon[169], but this is in its infancy—we will revisit this in sections 6.2 and 6.3.

A second mechanism for carbon extraction from the atmosphere-ocean system concerns expansion of the biosphere, most notably through reforestation. We have discussed this before in terms of expansion and contraction of the biosphere during ice-age cycles. In today's case, carbon extraction through biosphere expansion requires first that the industrial age's trend of net *de*forestation is reversed. Interestingly, this actually may have happened at around 2003. Between 2003 and 2014, net global vegetation increased by about 4 GtC (i.e., at an average rate of about 0.4 GtC per year), due to a lucky combination of increased rainfall on the savannahs of Australia, Africa, and South America, regrowth of forests on abandoned farmland in Russia and former Soviet republics, and massive tree-planting projects in China.[170] For an effective reforestation scenario, we would need to ensure that this new trend persists, and that we substantially increase it.

We can get a grip on the numbers involved in the total potential of reforestation by assessing an extreme scenario of 100 GtC extraction through reforestation over a period of 50 years[2,72]. Realistically, this will include a component of new growth, and a component of carbon storage into newly established or reinvigorated soils. Very recent work has suggested that the total potential for carbon storage through these mechanisms may even be up to 4 times larger than the 100 GtC mentioned before,[171] which indicates that the 100 GtC number we use is not unrealistic. Total reforestation and afforestation is impossible because of humanity's large-scale agricultural needs for food security. Our estimate of 100 GtC extraction in 50 years implies an average extraction rate of 2 GtC/y. Comparison with current annual emissions of almost 10 GtC immediately clarifies that even such draconian reforestation efforts would only be effective if combined with immediate emission reductions. In the absence of imposed emissions reductions, our major reforestation scenario would only reduce net emissions by 20%. Although that is a very useful contribution, it is also painfully obvious that we cannot hope to simply "plant our way out of the CO_2 problem."

A third potential mechanism focusses on the opposite of fossil fuel utilization, namely carbon burial. Effectively, this amounts to potential future fossil fuel creation, as fossil fuels are formed by burial of carbon in sediments. Low-oxygen to anoxic (that is, no oxygen available at all) aquatic environments are particularly effective at doing so; key examples are swamps on land, and anoxic marine or lake environments. Burial of land-plant matter in swamps typically leads to a potential of coal formation (sometimes with gas). Burial of algal matter in anoxic lakes or marine basins leads to potential oil and gas formation.

Burial of trees in swamps requires major growth of trees, and then for these to die and get buried in wetlands while other trees grow in their places. Given the lifetime of trees, this is a slow process that will—as a first step—require the reforestation that we already accounted for above; we should not account for it twice. Note also that humanity has been very effective at either cultivating or reclaiming (draining) swamps and wetlands, so major carbon burial in swamps and wetlands does not seem likely in the near future. Let's instead consider the alternative process of burial of algal matter in anoxic lakes or marine basins, to see how the carbon removal by such a process compares with today's CO_2 emissions by contrasting their relative impacts in crude terms.

Some of the best documented marine carbon burial events in recent geological history concern the formation of organic-rich layers known as sapropels in the eastern Mediterranean Sea (the entire basin to the East of Sicily). Each sapropel episode lasted about 4 to 10 thousand years, and they formed with great regularity throughout the last 15 million years. The "metronome" that determined their regular spacing is orbital forcing of climate. Almost every 22 thousand years, the African monsoon became intensified in response to climatic changes driven by orbital precession cycles.[172] During those times, the basin's circulation became impeded by monsoon fresh-water runoff from the Nile and other (now dry) rivers draining North Africa. The impeded circulation caused a strong reduction in oxygenation of the deep sea, so that the basin over time became anoxic, which favored burial of algal organic matter that sank from surface waters.[173,174] The organic matter burial led to deposition of organic-rich, black sediment layers that are immediately evident in sediment cores drilled from the seafloor; the sapropels.

The most recent of these layers is known as sapropel S1. S1 formed everywhere in the eastern Mediterranean below roughly 1000 m water depth, from about 10 to 6 thousand years ago. It contains up to 2% of organic carbon. Other sapropels lasted longer and were richer, such as S5 (128 to 121 thousand years ago) which lasted 7 thousand years and which contains up to an average of 8% of organic carbon. Older sapropels were recently found to have contributed significantly to the formation of natural gas that is commercially being extracted. In other words, the sapropels are functioning as so-called source-rocks.

So, how do our fossil-fuel-based CO_2 emissions compare with the CO_2 extraction when such source-rocks were formed? Or put in another way: if nature were to respond to climate change by making a large sea like the Mediterranean anoxic (which incidentally would kill almost everything that lives at depth), then would that perhaps offer any help with reducing CO_2 levels?

To find out, we need to know the rate at which carbon was buried when sapropels were formed. For S1, this has been carefully determined at 65 milligrams per square centimeter per thousand years.[175] The eastern Mediterranean has a surface area of 1.65 million square km, and for depths greater than 1000 m, the area is 0.85 million square km (8,500 trillion square centimeters)[176]. Thus, the burial amounted to about 0.55 GtC per thousand years for S1 (note: 1000 milligrams in a gram, 1000 grams in a kilogram, 1000 kg in a ton; one billion tons in a Gt). For the about 4 times richer S5, total carbon extraction may be estimated at $4 \times 0.55 = 2.2$ Gt per *thousand* years. For comparison, today's carbon emissions are almost 10 GtC per year. In other words, we find that total carbon removal from the atmosphere over 1,000 years in the case of the organic-rich S5 is equal to modern carbon emissions in only 80 days. It is important to take a moment to let that sink in, as it really puts the magnitude of our carbon emissions into perspective: *We emit in 80 days the same amount of carbon as nature locked away in a considerable oceanic basin over a period of 1,000 years.*

The above calculation was only for the eastern Mediterranean, a basin with well-known and relatively recent carbon burial at depth. But even though the eastern Mediterranean may be a big pond to swim across, it is minute in comparison with the vast world ocean. Let's hypothetically scale the Mediterranean calculation up to the entire world ocean, and see how much closer that result comes to the scale of modern carbon emissions.

The world ocean's surface area is about 361 million square kilometers. About 88% of the world ocean is deeper than 1000 m:[177] 318 million square kilometers. That is 374 times larger than the area over which the eastern Mediterranean is deeper than 1000 m. If we scale the S5 estimate up to a fully global value, assuming that the entire world ocean below 1000 m became an anoxic region of carbon burial, then we find a hypothetical maximum global marine burial of $374 \times 2.2 = 823$ GtC over 1,000 years, or 0.82 GtC per year. Note that ocean-wide anoxia has in reality happened only rarely in Earth's history, and as far as we can determine, such conditions have hardly ever reached a truly worldwide scale.[52] Therefore, our scenario gives a strong overestimate of the potential impact of carbon burial in sediments.

We find that— if the entire world ocean were to respond to climate change by going into an (unrealistic) anoxic deep-sea mode with intensive carbon burial—the carbon removal from the atmosphere by this massive, hypothetical process could only offset about 8% of the annual atmospheric carbon increase due to human-induced emissions. So, even if the ocean did its utmost to help us through carbon burial, it would not nearly be enough. In the meantime, however, the process would cause mass extinction of almost everything below 1000 m in the world ocean; not a very desirable prospect.

All that we have discussed to this point has outlined the limitations to what nature can do. We have seen that current CO_2 rise due to industrial age emissions is almost 100 times faster than the fastest natural rises recorded in the last million years, and roughly 10 to 30 times faster than the most dramatic CIEs. This goes some way to explaining why our hypothetical extreme scenario of natural carbon extraction still is more than 10 times too slow. Put simply, we have found that natural processes of carbon release and extraction typically operate at rates that reach only a few percent of the rates of carbon release by anthropogenic emissions.

We also saw that, as a potentially human-assisted natural process, reforestation holds more promise for carbon extraction. If optimally implemented, it might offset perhaps 20% of the current emissions rate, for some 50 years. Unfortunately, there is a realistic limit of about 100 GtC to the total extraction that is possible through reforestation without cutting into requirements for global food security. For scale, this total realistic amount of extraction would be negated by current emissions in about one decade. Hence, reforestation has great potential to help, but it falls far short of solving the problem.

Our discussion of weathering emphasized that it is an exceedingly slow process, and it is hard to estimate what its annual average impact may be. In any case, it is clear that even the combined sum of the three main natural carbon extraction processes would struggle to reach 30% of the current rate of carbon emissions. This simple assessment, using well-understood numbers, clearly indicates that there is no justification at all to thinking that natural processes might somehow be able to offset our industrial-age carbon emissions.

In all of recorded geological history, no carbon extraction has ever happened at anything close to a sufficiently high rate to do so; the necessary mechanisms simply do not exist. Given that nature on its own cannot even stop the annual CO_2 increase due to our emissions, there definitely is no chance that it could manage a reduction of atmospheric CO_2 levels on the timescale of a century or two that is relevant to society (Box 6.1).

6.2. REQUIREMENT FOR HUMAN INTERVENTION

None of the calculations made here gives anything more than rough ballpark estimates. But the overall implications are evident. And they remain so when using more sophisticated calculations.[2,72] The key implication therefore is an urgent need for dramatic emission reductions.

The key question is *How strong should emission reduction be to stabilize (that is, to stop the increase of) atmospheric CO_2 levels?* We have seen that nothing in nature can offset more than about a third of the annual rate of emissions. Moreover, global ocean anoxia is not exactly desirable because it involves the extermination of virtually everything that lives below about 1000 m depth, so it is better not to count on that as part of a solution.

Box 6.1.

Our discussion should not be taken to imply that natural processes cannot remove the CO_2 at all. On the contrary, the short-term impact of the natural processes may be too small to help on timescales of centuries, but the long-lasting nature of the processes means that they will eventually manage to reduce the CO_2 levels if given enough time (thousands to hundreds of thousands of years). This will be predominantly through chemical weathering on land and carbonate compensation within the ocean. Thus, nature will eventually clean up our mess, but at timescales that are far beyond relevance to society. For the best documented natural burp of external carbon into the hydrosphere-biosphere-atmosphere system (the PETM), the clean-up took some 200,000 years.

Ignoring global ocean anoxia leaves only large-scale reforestation as a somewhat viable (human-assisted) natural carbon removal process, for a period of about five decades. If reforestation were optimized, then CO_2 stabilization at current levels would require urgent, preferably immediate, emission reduction to roughly 2 GtC per year (that is, to only 20% of the current rate).

Once reforestation is complete after five decades or so, the biosphere will no longer offer a sizeable potential for longer-term net carbon removal, lest we risk damaging global food security. By that time, we need to be ready to run our entire society with zero net emissions, in spite of population growth. Any remaining emissions from cement manufacture and fossil carbon use from that time will need to be artificially removed from the atmosphere, or prevented from entering the atmosphere in the first place, to achieve zero net emissions.

Thinking through the matter further now reveals that just reduction of emissions is not enough because zero emissions alone will not suffice to keep global warming well below the Paris agreement's 2 °C target. This turns the focus onto policy changes that will facilitate the development of capacity to draw carbon out of the atmosphere-ocean system[178] toward a long-term target of CO_2 levels at around 350 ppm.

To make atmospheric CO_2 levels drop, large-scale (human-controlled) net carbon capture and removal will be essential. Those efforts would be *in addition to* all measures discussed previously—including the entire global reforestation potential—to achieve zero net emissions (which would only suffice to keep CO_2 levels constant).

The geological perspective we have developed through this book has shown that there is an oceanic sting in the tail where reducing atmospheric CO_2 levels is concerned. The ocean has taken up just under half of our emissions, in a process called equilibration between atmospheric and oceanic levels. Thus, the atmospheric CO_2 increase since pre-industrial times has been only half as large as it could have been, if the ocean had not equilibrated with the atmosphere. Unfortunately, this knife cuts both ways—if we reduce atmospheric CO_2 levels, then equilibration means that the levels in the ocean will also reduce. This will be achieved by returning CO_2 to the atmosphere. The atmosphere/ocean ratio of roughly $1/1^2$ or $1.5/1.0^{71}$ again applies for this reversed process. In consequence, our target 50 ppm reduction of atmospheric CO_2 (from about 400 to 350 ppm) will require capture and removal of 1.7 to 2 times more carbon than we would think on the basis of the atmospheric amount alone. Given that 1 ppm of CO_2 is equivalent to about 2.12 GtC, a reduction of 50 ppm thus implies a need for net carbon capture and removal of 180 to 212 GtC—let's say 195 GtC for ease of calculation. In sheer quantity, this is twice as much as that involved in the total potential global reforestation. But remember that our calculations used the reforestation potential already (along with 80% emissions reduction) to stabilize CO_2 levels, so we cannot use it again. Therefore, the need for 195 GtC extra removal mentioned here to achieve CO_2 *reduction*, is *additional* to the reforestation and 80% emissions reduction. This is a massive technological and environmental challenge.

We can try to get a grip on the scale of this challenge by focusing on one of the more promising ways under consideration for large-scale CO_2 removal. It is the process of artificial weathering of magnesium-rich minerals (notably olivine). In a theoretical best case scenario, one kilogram of olivine is needed to remove one kilogram of CO_2.[179] Because the atomic mass of carbon is 12, and the combined atomic masses in CO_2 add up to 44, our 195 Gt of carbon is equivalent to about 715 Gt of CO_2. Removal of such a mass of CO_2 from the atmosphere would require 715 Gt of olivine in the best case scenario.

What sort of mining operations would be needed to gain 715 Gt of olivine? Olivine weighs about 3.5 g per cubic centimeter, or 3.5 tons per cubic meter, so 715 billion tons of olivine equate to about 204 billion cubic meters of olivine. Olivine-weathering pioneer Olaf Schuiling pointed out to me that the most effective approach to mining it is by focusing on rocks called dunite, which contain more than 90% of olivine.[180] Thus, we would need to excavate and process 204 to 227 billion cubic meters of dunite.

A sense of scale for this can be gained from considering the surface area of England, which is about 130 billion square meters. To mine the quantity of olivine needed in our artificially stimulated weathering experiment, we would need to remove of a layer of some 1.6 m of pure dunite over an area the size of England to support our best case scenario for CO_2 reduction by artificial weathering. Alternatively, we can gauge the scale of the operation by comparison with today's deepest and almost largest excavated hole in the world: the Bingham Canyon Mine, also known as the Kennecott Copper Mine (Utah, USA). It is about 1000 m deep and 4000 m wide. Its excavated volume amounted to some 1.25 billion cubic meters in 2008 already[181]. We find that our theoretical dunite mine would need to be 163 to 182 times larger than the Bingham Canyon Mine. Mining on this sort of scale may not be impossible, but it certainly poses a gargantuan challenge. And remember that—to not defeat the purpose of the exercise—this titanic mining effort would need to do all its extracting, processing, and transportation with zero net carbon emissions.

The outlined mega-challenge is in addition to efforts toward emissions reduction. These efforts include alternative energy generation, electrification of transport and other infrastructure combined with carbon capture at source in central electricity generation, geothermal heating, changing societal attitudes and habits, and many others. In the final section I will focus on carbon removal solutions.

6.3. HUMAN INTERVENTION IN CARBON REMOVAL

As mentioned, the amount of carbon capture and removal from the climate system is of almost unimaginable proportions. In the end, we will need every single option that we can develop, and that has the least environmental impact and the greatest societal acceptability.

Realistically, such a task can only be contemplated by enlisting the services of the wider Earth System. Somehow, we need carbon to be removed from the active climate system, and stored in rocks and/or soils. A vegetation (biosphere) expansion that is thereafter kept in good condition offers a one-time carbon-removal process. Restoration and enrichment of soils creates a longer-term storage option. It may not be entirely closed off from the active climate system, but at least it offers respite of the order of several centuries or more. The same is true for transport of captured carbon—via plankton blooms or other processes—into the deeper ocean. Eventually, most of that carbon will resurface. We should therefore mainly think of these processes as means for buying us a few centuries, which we ought to then use for sorting out more permanent solutions. In contrast, storage in the form of rocks is ideal, because it creates external carbon that is truly away from exchange with the active climate system.

But the first problem is how to get the CO_2 out of the atmosphere. Technologically, we know how to do that on smaller scales, given that it's been done for a long time already in submarines and inhabited spacecraft. The issue lies with upscaling the direct air capturing (DAC) technology to the enormous facilities that will be needed all around the world to make a difference sucking diffuse CO_2 from the atmosphere, and with how to power these facilities. This is why carbon capture is mostly trialed on power plants; they produce enough power and their emissions are easily accessible in the form of concentrated point sources. When this carbon capture is linked to storage of the captured products, for example in depleted oil and gas fields, the technology is often described as carbon capture and storage (CCS). CCS underpins "clean coal," which consists of burning coal in more efficient ways than traditionally done (to improve its energy yield), and to then capture the emitted CO_2 at source and store it[182].

But "clean coal" is riddled with technological and logistical complications, and cannot be realistically done by upgrading existing power plants. It therefore involves the construction of new coal power plants and thus the prolongation of the world's fossil fuel addiction, to the detriment of alternative energy implementation. Moreover, "clean coal" does not address the host of other heavy pollutants that accompany the use of coal for energy. CCS has also been proposed for larger industrial complexes such as ports, which are more diffuse sources than power plants, but none has reached a stage anywhere close to implementation (yet).

Bioenergy with carbon capture and storage (BECCS) is another proposed way of working toward negative emissions[183]. In this approach, plant/tree biomass is grown to produce fuel for bioenergy power plants; this growth draws CO_2 from the atmosphere. The biomass is then burned to produce energy, and the carbon emissions are captured at source. The captured carbon then needs to be stored in some way to remove it from the active climate system. Thus, BECCS would drive net carbon removal from the atmosphere.

One issue again lies with up-scaling BECCS. If it is to make a difference in the enormous carbon removal requirement from the climate system, BECCS would need to be applied on enormous scales, and in addition rapid technological improvements are essential[184]. And it will impose strong demands for water and fertile lands for growing its biomass[185], which will seriously interfere with agricultural requirements for food security and/or natural ecosystems. A very similar conflict happened when bioethanol became fashionable for fueling cars. With a global transition to BECCS, this conflict would be driven onto a different scale altogether. Another issue with BECCS is the need to store the captured carbon. One of the main proposals, injection into certain rocks (like basalt), leads to the formation of magnesium-carbonate minerals, which are stable and lock the carbon away. However, this requires BECCS plants to be on top of, or close to, such rocks. And that puts some limitations to their spatial distribution. Also, hydro-fracturing (fracking) of basement rocks will likely be needed to facilitate the injection, which may be a matter of significant concern among neighboring populations. Neither problem is unsurmountable, but both are difficult and restrictive nonetheless.

These perceived limitations to technological solutions, mostly to do with scaling them up to sufficient sizes for making a significant impact on atmospheric CO_2 levels, drive a burgeoning interest in marshalling the Earth System (carbon cycle) itself for both the capturing and storage of carbon[72,186] (Figure 4.1). *"Impossible,"* you may say, but consider this: it's exactly what we have done in the opposite direction with our fossil fuel burning. The natural oxidation of organic-rich sediments and fossil fuels was a very small flux term before we got started, but once we got started we quickly magnified it through industrial processes. As time went by, we got better and better at it, and now we emit a massive 10 GtC each year (Figure 1.2).

The problem we're facing now is a bit different, in that the main carbon removal mechanisms are more diffuse in nature—they are not so nicely concentrated as fossil fuels—and that they also have very small flux terms. For example, the net weathering flux of carbon removal from the active climate system is only 0.3 GtC/y (Figure 4.1). If we want to remove 195 GtC of carbon in the about 80 years remaining in this century, we would somehow need to accelerate the global weathering processes by 8 times. This is a daunting task, and we have seen that it will require gigantic amounts of suitable minerals. But there may be clever ways of doing it.

For example, a much-voiced idea is to mine, grind, and then spread rock dust onto agricultural land[187]. The rock dust, which can be from basalt-like rocks that are widespread around the world, also contains many nutrient-rich minerals. As chemical weathering occurs, it consumes CO_2, and the nutrients are released into the soil. The weathered rock in addition causes the alkalinity of draining waters to increase, and this in turn causes increased CO_2 invasion into the waters[188]. Because of its nutrient release, the rock dust acts to fertilize the soil, which helps to increase crop yields and reduces the need for artificial fertilizers, although it's important to keep rock-dust/olivine doses within limits that avoid imbalances in plant nutrition.[189] Overall, the soil gets enriched and holds more carbon, crop yields increase, and carbon is extracted from the atmosphere. Richer soils moreover may retain water better than depleted soils, helping to alleviate the effects of occasional droughts, and/or reduce the need for tapping into precious water resources for irrigation. Furthermore, it seems that ground stability may be improved.[190]

The crops produced with rock dust fertilization could be food, or biomass for BECCS plants that would lead to further carbon extraction. Or the rock dust fertilization could be used in places for reforestation/afforestation, restoring natural ecosystems through soil restoration efforts and planting of native vegetation. Especially if wetlands are re-developed this way, even more carbon extraction and storage in those natural environments will occur.

All this may sound like utopia, and it's been tested only at small scales. Once it is done for real on very large scales and over very long periods of time, the apparent co-benefits may work out differently than portrayed here, and the efficiencies of the various processes may not be as great as some models suggest (for example, soils may "saturate" and stop taking up carbon). Consequently, urgent research and medium to large-scale testing is needed. The same is true for a range of other ideas; for an excellent overview, see Project Drawdown.[191] In many cases, only model results exist, which rely on short-term tests in laboratories or greenhouses.

Gradually, the climate of research funding and major corporate sponsorship around the world is waking up to the urgency of working out the problems with carbon extraction from the climate system, so that we can identify the promising approaches and weed out the weak ones. Greater urgency is needed, though: the scale of the problem is such that we will eventually need to implement every single workable solution available.

[7]

SUMMARY

Several independent series of observations demonstrate that there has been about 1°C of warming since the start of the industrial revolution. We discussed that there is some variability in solar output, and that these variations may be recognized in records of past climate, but also that solar variability can only account for warming by 0.1°C to an unlikely maximum of 0.35°C since the end of the Little Ice Age. Based on energy balance considerations, we have found that our emissions of external carbon are the main culprit.

In response to this disturbance of the energy balance, the climate simply has to change toward a warmer Pliocene-like state, even if we could manage to stabilize CO_2 at its current level of about 400 ppm. From discussion of several slowly adjusting processes within the climate system, we now understand that it will take, from the beginning, several centuries to approximate the full Pliocene-like warming. But we are almost two centuries down the road, and warming to date already amounts to about 1°C. The slow components in the climate system will cause continuing warming by another 1°C or so. In other words, we are already committed to further warming, even if we managed to make the massive jump to a zero-emissions society from today and thus stabilize CO_2 levels.

The urgency of slashing back the current level of annual emissions (10 GtC) cannot be overstated. Every year of inaction brings us closer to the inevitability of a future climate that will exceed even the warm Pliocene state, with global temperatures at least 2 or 3°C higher than the pre-industrial level. If we allow ourselves to reach the Paris Climate Conference's agreed maximum of 2°C warming by the year 2100, then the further commitment over coming centuries would take us toward 4°C, even if we achieved zero emissions by 2100. That is considerably warmer than during the Pliocene.

We have seen that the consequences are grave. Progression toward a Pliocene-like climate state will be accompanied by continued migration of global and regional climate zones and by intensification of the evaporation and precipitation cycle, placing many areas at risk of increasing extremes, including aridity, flooding, and lethal heat. Meanwhile, as the large continental ice sheets adjust to the changed conditions and cause further warming themselves through the ice-albedo feedback, there will be an unstoppable rise in sea level. This adjustment will continue over many centuries because of the long response timescales of large ice sheets. We evaluated geological data for an indication of how severe this impact may become over time, even if we would stabilize CO_2 at 400 ppm. Pliocene sea levels stood at least 9 m and, according to most estimates, between 12 and 32 m above the present level. The geological data also show that rates of sea-level rise above the present level are likely to reach at least 1 m per century and even (much) more after 2100.

Having determined that nature cannot offset more than perhaps a third of our annual emissions, we know for certain that CO_2 levels will continue to rise unless we drastically slash the rate of emissions. And even if we manage to stabilize CO_2 levels, we still need to seek ways of reducing them back toward 350 ppm to restore long-term climate to "safe" conditions. There is nothing in nature that can do this for us on timescales relevant to society. Nature can only do it over hundreds of thousands of years.

We assessed that the remedial measures that we need to develop are of a truly unprecedented scale in human history. The onus is on us— right here and right now—to find engineering and/or Earth System-based solutions. This represents a technological and logistic challenge like no other before. But consider this: humans have always excelled at overcoming extremely unfavorable odds. We should see this challenge as the impetus for launching into a much-needed transition into sustainable operation of society. This is our chance to civilize civilization.

[8]

EPILOGUE

There can be no doubt that the discovery and widespread use of fossil fuels has driven massive economical and societal developments. It started with coal. Although oil and gas became important thereafter, coal never really lost its position of prominence, although the tide may have turned in very recent years (Figure 1.2). The use of fossil fuels is familiar to virtually everybody on the planet, and a massive global corporate infrastructure exists to ensure its supply and distribution. In addition, petrochemical products are essential base materials for much of the manufacturing industry (e.g., plastics, synthetics, solvents, etc.). Virtually all of humanity has thus become thoroughly dependent on fossil fuels and other petrochemical products. We feel in our comfort zone with them.

The very closely interwoven relationship between the modern way of life and fossil fuels and other petrochemicals, the emotive comfort-zone issue, and direct financial interests are key to the strong reactions that commonly follow whenever external carbon emissions are identified as an urgent problem. I hope that this book has provided you with a deeper understanding of the concepts and observations behind the concerns expressed by researchers. The role of CO_2 in climate could perhaps be defined a little more sharply in the fine details, but the big lines are clear. And the big lines are all we need to appreciate where we stand today. Using just a few simple numbers about the potential consequences of the modern CO_2 levels, with examples from the past, and equally simple numbers about the sheer quantities of carbon involved, everyone can get their heads around the scale of the problem.

By similarly working out the scale of the required remedial measures in basic terms, we have seen that our most advisable course of action is to create a zero emissions society at the very soonest. Because there has been inaction for so long already, positive action on a truly global scale has now become a matter of the greatest urgency. This zero emission society is critical for stopping the unnaturally rapid rise in CO_2 levels and stabilizing them to a level as close to today's 400 ppm as we can. In parallel, we need to launch a major push on development and implementation of technologies for active CO_2 reduction.

There will be major costs involved. But humanity is where it is because it has never turned away from major challenges. It has always overcome these challenges by large-scale innovation. Such innovation in turn leads to new exploitable opportunities and offshoots in technology and engineering, which will give considerable returns on investment along the way. Given that there is no way in which nature can clean up the environment for us on timescales that are relevant to society, we simply have no choice. We are on our own: it is entirely up to us, and us alone, to clean up after ourselves.

If we accept that responsibility, then it is critical that we do not passively wait for the required innovations to occur by chance, but that we establish and nurture an intellectual environment and can-do culture that actively encourages and drives innovation and its implementation. The costs of inaction are not just financial, but also—and more importantly— humanitarian and environmental. Inaction will increasingly impact on water and food resources on national to international scales and will therefore become a liability; it will be (or arguably already is)[192] a likely cause for future hardship and conflict. This won't be a new phenomenon—in the past, even relatively modest natural climate changes are thought to have been important contributors to large-scale societal upheavals on regional scales.[193,194,195]

There are some broader issues at stake, too.

First is the issue of our dealing with *any* kind of pollution. The carbon problem is not the only form of pollution. Other examples include illegal dumping, mindless littering, turning a blind eye to detrimental practices, and many other examples. With 7+ billion people on the planet already, survivability requires a focus on avoiding pollution and degradation of the one and only habitat we have available to us. This reason gives just as much justification to develop green energy as the need to reduce carbon emissions.

Second is the issue of much-voiced objections to green energy development for aesthetic or contrived health reasons. It is not as if the fossil-fuel industry has no large-scale infrastructure with drawbacks like being unsightly, light-polluting, noisy, stinky, leaky, and otherwise unpleasant. All major developments have such drawbacks, but we have learned to live with them, or—more cynically—have often left the less privileged to live with them. A more balanced view of the available alternatives is needed. People objecting to green energy developments usually are not so keen on the alternatives (fossil fuels or nuclear) that they would swap a planned solar or wind farm for a major oil refinery, coal-burning power station, or nuclear reactor in their backyard. Yes, a transition to a sustainable future will involve some tough choices, but tough choices are already being made in today's industrialized society, and the existing choices clearly are not sustainable. To survive, we must develop green, sustainable energy.

Third is the issue of resistance in many nations to positively emphasize development of the knowledge base and infrastructure for a strong portfolio of alternative energy generation. Such a portfolio would reduce the reliance on fossil fuels and—through energy diversity—offer greater energy security and self-sufficiency. Taking the extra step toward clean air and reduced pollution would also bring major benefits to population health, similar to the past improvements driven by the transition from coal fired energy to relatively cleaner oil and gas-fired energy in major cities around the world. Development of the knowledge base and infrastructure for a strong portfolio of alternative energy generation and carbon removal may require investment, but it is also likely to attract overseas investment and business opportunities. Moreover, it offers great potential for the nations involved to become an international hub for exporting technology and knowledge, with major paybacks in employment and financial gain. These benefits add to the returns gained from enduring reduction in the costs for health care owing to reduced pollution.

Great opportunities lie open in places such as Australia, which has more space per person than almost any other nation—large swathes of it replete with solar, wind, wave, and tidal power—and which already has the requisite education system, pioneering mentality, and financial backbone. It might re-invent itself as a showcase and true global leader in energy self-sufficiency, a global driver in technology innovation, and exporter of green technology and knowledge. Such innovations could extend to programs designed toward carbon extraction, such as large-scale soil restoration and reforestation/afforestation in currently depleted, and consequently underexploited, regions. Many other nations have similar opportunities.

But seizing opportunity requires the courage and foresight for any nation to step out of the fossil-fuel comfort zone and invest in order to take the lead in a positive advance toward a cleaner society. Let's recognize and seize the opportunity and act with vision for the future, instead of sticking to a comfort zone that dates back one or two centuries and will not last.

NOTES

Chapter 1

1. Knutti, R., Rogelj, J., Sedláček, J., and Fisher, E.M., A scientific critique of the two-degree climate change target. *Nature Geoscience, 9,* 13–18, 2016.
2. Hansen, J., Kharecha, P., Sato, M., Masson-Delmotte, V., Ackerman, F., Beerling, D., Hearty, P.J., Hoegh-Guldberg, O., Hsu, S.-L., Parmesan, C., Rockstrom, J., Rohling, E.J., Sachs, J., Smith, P., Steffen, K., Van Susteren, L., von Schuckmann, K., and Zachos, J.C., Assessing "dangerous climate change": Required reduction of carbon emissions to protect young people, future generations and nature. *PLoS ONE, 8,* e81648, doi:10.1371/journal.pone.0081648, 2013.
3. Mann, M.E., Earth will cross the climate danger threshold by 2036. *Scientific American,* April 1, 2014. http://www.scientificamerican.com/article/earth-will-cross-the-climate-danger-threshold-by-2036/
4. For example: Japan Meteorological Agency; UK Hadley Centre; US National Oceanic and Atmospheric Administration (NOAA) and National Climatic Data Center (NCDC); US NASA-GISS (Goddard Institute for Space Studies); US Berkeley Earth.
5. http://ds.data.jma.go.jp/tcc/tcc/products/gwp/temp/ann_wld.html

Chapter 2

6. Parrenin, F., Barnola, J.M., Beer, J., Blunier, T., Castellano, E., Chapellaz, J., Dreyfus, G., Fischer, H., Fujita, S., Jouzel, J., Kawamura, K.,

Lemieux-Dudon, B., Loulergue, L., Masson-Delmotte, V., Narcisi, B., Petit, J.R., Raisbeck, G., Raynaud, D., Ruth, U., Schwander, J., Severi, M., Spanhi, R., Steffensen, J.P., Svensson, A., Udisti, R., Waelbroeck, C., and Wolff, E., The EDC3 chronology for the EPICA Dome C ice core. *Climate of the Past*, 3, 485–497, 2007.

7. Siegenthaler, U., Stocker, T.F., Monnin, E., Lüthi, D., Schwander, J., Stauffer, B., Raynaud, D., Barnola, J.M., Fischer, H., Masson-Delmotte, V., and Jouzel, J., Stable carbon cycle–climate relationship during the Late Pleistocene. *Science*, 310, 1313–1317, 2005.

8. Lüthi, D., Le Floch, M., Bereiter, B., Blunier, T., Barnola, J.M., Siegenthaler, U., Raynaud, D., Jouzel, J., Fischer, H., Kawamura, K., and Stocker, T., High-resolution carbon dioxide concentration record 650,000–800,000 years before present. *Nature*, 453, 379–382, 2008.

9. Spahni, R., Chapellaz, J., Stocker, T.F., Loulergue, L., Hausammann, G., Kawamura, K., Flückiger, J., Schwander, J., Raynaud, D., Masson-Delmotte, V., and Jouzel, J., Atmospheric methane and nitrous oxide of the Late Pleistocene from Antarctic ice cores. *Science*, 310, 1317–1321, 2005.

10. Loulergue, L., Schilt, A., Spanhi, R., Masson-Delmotte, V., Blunier, T., Lemieux, B., Barnola, J.M., Raynaud, D., Stocker, T.F., and Chapellaz, J., Orbital and millennial-scale features of atmospheric CH$_4$ over the past 800,000 years. *Nature*, 453, 383–386, 2008.

11. Cheng, H., Edwards, R.L., Broecker, W.S., Denton, G.H., Kong, X., Wang, Y., Zhang, R., and Wang, X., Ice age terminations. *Science*, 326, 248–252 (2009).

12. Cheng, H., Zhang, P.Z., Spötl, C., Edwards, R.L., Cai, Y.J., Zhang, D.Z., Sang, W.C., Tan, M., and An, Z.S., The climatic cyclicity in semiarid-arid central Asia over the past 500,000 years. *Geophysical Research Letters*, 39, L01705, doi:10.1029/2011GL050202, 2012.

13. Bar-Matthews, M., Ayalon, A., Gilmour, M., Matthews, A., and Hawkesworth, C.J., Sea–land oxygen isotopic relationships from planktonic foraminifera and speleothems in the Eastern Mediterranean region and their implication for paleorainfall during interglacial intervals. *Geochimica et Cosmochimica Acta*, 67, 3181–3199, 2003.

14. Esper, J., Frank, D.C., Timonen, M., Zorita, E., Wilson, R.J.S., Luterbacher, J., Hozkämper, S., Fischer, N., Wagner, S., Nievergelt, D., Verstege, A., and Büntgen, U., Orbital forcing of tree-ring data. *Nature Climate Change*, 2, 862–866, 2012.

15. Allen, J.R., Brandt, U., Brauer, A., Hubberten, H.W., Huntley, B., Keller, J., Kraml., M., Mackensen, A., Mingram, J., Negendank, J., Nowaczuk, N.R., Oberhänsli, H., Watts, W.A., Wulf, S., and Zolitschka, B., Rapid environmental changes in southern Europe during the last glacial period. *Nature*, 400, 740–743, 1999.

16. Zolitschka, B., Brauer, A., Negendank, J., Stockhausen, H., and Lang, A., Annually dated late Weichselian continental palaeoclimate record from the Eifel, Germany. *Geology*, 28, 783–786, 2000.
17. Torfstein, A., Goldstein, S.L., Kagan, E., and Stein, M., Integrated multi-site U-Th chronology of the last glacial Lake Lisan. *Geochimica et Cosmochimica Acta*, 104, 210–231, 2013.
18. Torfstein, A., Goldstein, S.L., Stein, M., and Enzel, Y., Impacts of abrupt climate changes in the Levant from last glacial Dead Sea levels. *Quaternary Science Reviews*, 69, 1–7, 2013.
19. Rohling, E.J., Quantitative assessment of glacial fluctuations in the level of Lake Lisan, Dead Sea rift. *Quaternary Science Reviews*, 70, 63–72, 2013.
20. Broecker, W.S., and Denton, G.H., What drives glacial cycles? *Scientific American*, 262, 48–56, 1990.
21. Kuhlemann, J., Rohling, E.J., Kumrei, I., Kubik, P., Ivy-Ochs, S., and Kucera, M., Regional synthesis of Mediterranean atmospheric circulation during the Last Glacial Maximum. *Science*, 321, 1338–1340, doi:10.1126/science.1157638, 2008.
22. Mayewski, P.A., Rohling, E.J., Stager, J.C., Karlén, W., Maasch, K., Meeker, L.D., Meyerson, E., Gasse, F., Van Kreveld, S., Holmgren, K., Lee-Thorp. J., Rosqvist, G., Rack, F., Staubwasser, M., Schneider, R.R., and Steig, E., Holocene climate variability. *Quaternary Research*, 62, 243–255, 2004.
23. Kürschner, W.M., van der Burgh, J., Visscher, H., and Dilcher, D.L., Oak leaves as biosensors of late Neogene and early Pleistocene paleoatmospheric CO_2 concentrations. *Marine Micropaleontology*, 27, 299–312, 1996.
24. Retallack, G.J., A 300-million-year record of atmospheric carbon dioxide from fossil plant cuticles. *Nature*, 411, 287–290, 2001.
25. Schefuß, E., Schouten, S., and Schneider, R.R., Climatic controls on central African hydrology during the past 20,000 years. *Nature*, 437, 1003–1006, 2005.
26. Tierney, J.E., Russell, J.M., Huang, Y., Sinninghe-Damsté, J.S., Hopmans, E.C., and Cohen, A.S., Northern hemisphere controls on tropical southeast African climate during the past 60,000 years. *Science*, 322, 252–255, 2008.
27. Tierney, J.E., Lewis, S.C., Cook, B.I., LeGrande, A.K., and Schmidt, G.A., Model, proxy and isotopic perspectives on the East African Humid Period. *Earth and Planetary Science Letters*, 307, 103–112, 2011.
28. Lourens, L.J., Antonarakou, A., Hilgen, F.J., Van Hoof, A.A.M., Vergnaud-Grazzini, C., and Zachariasse, W.J., Evaluation of the Plio-Pleistocene astronomical time-scale. *Paleoceanography*, 11, 391–413, 1996.
29. Lourens, L.J., Wehausen, R., and Brumsack, H.J., Geological constraints on tidal dissipation and dynamical ellipticity of the earth over the past three million years. *Nature*, 409, 1029–1033, 2001.
30. Wade, B.S., Pearson, P.N., Berggren, W.A., and Pälike, H., Review and revision of Cenozoic tropical planktonic foraminiferal biostratigraphy and calibration

to the Geomagnetic Polarity and Astronomical Time Scale. *Earth Science Reviews, 104*, 111–142, 2011.

31. Pälike, H., Laskar, J., and Shackleton, N.J., Geological constraints on the chaotic diffusion of the Solar System. *Geology, 32*, 929–932, 2004.

32. Pälike, H., and Shackleton, N.J., Constraints on astronomical parameters from the geological record for the last 25 My. *Earth and Planetary Science Letters, 182*, 1–14, 2000.

33. Rohling, E.J., and Cooke, S., Stable oxygen and carbon isotope ratios in foraminiferal carbonate, chapter 14 in B.K. Sen Gupta (ed.), *Modern Foraminifera*. Kluwer Academic, Dordrecht, The Netherlands, pp. 239–258, 1999.

34. Yu, J., Anderson, R.F., and Rohling, E.J., Deep ocean carbonate chemistry and glacial-interglacial atmospheric CO_2 changes. *Oceanography, 27*, 16–25, 2014.

35. Broecker, W.S., and Peng, T.-H., The role of $CaCO_3$ compensation in the glacial to interglacial atmospheric CO_2 change. *Global Biogeochemical Cycles, 1*, 15–29, 1987.

36. Zachos, J.C., Röhl, U., Schellenberg, S.A., Sluijs, A., Hodell, D.A., Kelly, D.C., Thomas, E., Nicolo, M., Raffi, I., Lourens, L.J., McCarren, H., and Kroon, D., Rapid acidification of the ocean during the Paleocene-Eocene Thermal Maximum. *Science, 308*, 1611–1615, 2005.

37. Zeebe, R.E., Zachos J.C., and Dickens, G.R., Carbon dioxide forcing alone insufficient to explain Palaeocene–Eocene Thermal Maximum warming. *Nature Geoscience, 2*, 576–580, 2009.

38. Van Andel, T.H., Heath, G.R., and Moore, T.C. Jr., Cenozoic history and paleoceanography of the central equatorial Pacific Ocean: a regional synthesis of Deep Sea Drilling Project data. *Geological Society of America, 143*, 1–134, 1975.

39. Pälike, H., Lyle, M.W., Nishi, H., Raffi, I., Ridgwell, A., Gamage, K., Klaus, A., Acton, G., Anderson, L., Backman, J., Baldauf, J., Beltran, C., Bohaty, S.M., Bown, P., Busch, W., Channell, J.E.T., Chun, C.O.J., Delaney, M., Dewangan, P., Dunkley Jones, T, Edgar, K.M., Evans, H., Fitch, P., Foster, G.L., Gussone, N., Hasegawa, H., Hathorne, E.C., Hayashi, H., Herrle, J.O., Holbourn, A., Hovan, S., Hyeong, K., Iijima, K., Ito, T., Kamikuri, S., Kimoto, K., Kuroda, J., Leon-Rodriguez, L., Malinverno, A., Moore, T.C. Jr., Murphy, B.H., Murphy, D.P., Nakamura, H., Ogane, K., Ohneiser, C., Richter, C., Robinson, R., Rohling, E.J., Romero, O., Sawada, K., Scher, H., Schneider, L., Sluijs, A., Takata, H., Tian, J., Tsujimoto, A., Wade, B.S., Westerhold, T., Wilkens, R., Williams, T., Wilson, P.A., Yamamoto, Y., Yamamoto, S., Yamazaki, T., and Zeebe, R.E., A Cenozoic record of the equatorial Pacific carbonate compensation depth. *Nature, 488*, 609–614, 2012.

40. Elderfield, H., Ferretti, P., Greaves, M., Crowhurst, S., Mccave, I.N., Hodell, D., and Piotrowski, A.M., Evolution of ocean temperature and ice volume through the Mid- Pleistocene Climate Transition. *Science, 337*, 704–709, 2012.

41. Lear, C.H., Bailey, T.R., Pearson, P.N., Coxall, H.K., and Rosenthal, Y., Cooling and ice growth across the Eocene-Oligocene transition. *Geology, 36,* 251–254, 2008.
42. Rohling, E.J., Foster, G.L., Grant, K.M., Marino, G., Roberts, A.P., Tamisiea, M.E., and Williams, F., Sea-level and deep-sea-temperature variability over the past 5.3 million years. *Nature, 508,* 477–482, 2014.
43. Grant, K.M., Rohling, E.J., Bar-Matthews, M., Ayalon, A., Medina-Elizalde, M., Bronk Ramsey, C., Satow, C., and Roberts, A.P., Rapid coupling between ice volume and polar temperature over the past 150 kyr. *Nature, 491,* 744–747, 2012.
44. Rohling, E.J., Fenton, M., Jorissen, F.J., Bertrand, P., Ganssen, G., and Caulet, J.P., Magnitudes of sea-level lowstands of the past 500,000 years. *Nature, 394,* 162–165, 1998.
45. Siddall, M., Rohling, E.J., Almogi-Labin, A., Hemleben, Ch., Meischner, D., Schmeltzer, I., and Smeed, D.A., Sea-level fluctuations during the last glacial cycle. *Nature, 423,* 853–858, 2003.
46. Shennan, I., Long, A.J., and Horton, B.P. (eds.), *Handbook of Sea-Level Research.* American Geophysical Union, John Wiley & Sons, 600 pp., ISBN: 978-1-118-45258-5, 2015.
47. http://strata.uga.edu/sequence/index.html
48. http://www.searchanddiscovery.com/pdfz/documents/2010/40594snedden/ndx_snedden.pdf.html

Chapter 3

49. Pollack, H.N., Hurter, S.J., and Johnson, J.R., Heat flow from the Earth's interior: analysis of the global data set. *Reviews of Geophysics, 31,* 267–280, 1993.
50. The KamLAND Collaboration, Partial radiogenic heat model for Earth revealed by geoneutrino measurements. *Nature Geoscience, 4,* 647–651, 2011.
51. Rohling, E.J., Medina-Elizalde, M., Shepherd, J.G., Siddall, M., and Stanford, J.D., Sea surface and high-latitude temperature sensitivity to radiative forcing of climate over several glacial cycles. *Journal of Climate, 25,* 1635–1656, 2012.
52. Rohling, E.J., *The oceans: a deep history.* Princeton University Press, 272 pp., 2017. ISBN9781400888665
53. Tyndall, J., *Contributions to Molecular Physics in the Domain of Radiant Heat*—A series of memoirs published in the "Philosophical Transactions and Philosophical Magazine," with additions. Longmans, Green, and Co., London, 450 pp., 1872.

54. Arrhenius, S., On the influence of carbonic acid in the air upon the temperature of the ground. *The London, Edinburgh, and Dublin Philosophical Magazine and Journal of Science, 5*, 237–275, 1896.

55. Feulner, G., The faint young Sun problem. *Reviews of Geophysics, 50*, RG2006, 10.1029/2011RG000375, 29 pp., 2012.

56. http://www.engineeringtoolbox.com/water-vapor-saturation-pressure-air-d_689.html

57. Pierrehumbert, R.T., Thermostats, radiator fins, and the local runaway greenhouse. *Journal of the Atmospheric Sciences, 52*, 1784–1806, 1995 (especially figure 2 in that study).

58. Royer, D.L., Pagani, M., and Beerling, D.J. Geobiological constraints on Earth system sensitivity to CO_2 during the Cretaceous and Cenozoic. *Geobiology, 10*, 298–310, 2012.

59. PALAEOSENS Project Members (Rohling, E.J., Sluijs, A., Dijkstra, H.A., Köhler, P., van de Wal, R.S.W., von der Heydt, A.S., Beerling, D., Berger, A., Bijl, P.K., Crucifix, M., deConto, R., Drijfhout, S.S., Fedorov, A., Foster, G., Ganopolski, A., Hansen, J., Hönisch, B., Hooghiemstra, H., Huber, M., Huybers, P., Knutti, R., Lea, D.W., Lourens, L.J., Lunt, D., Masson-Demotte, V., Medina-Elizalde, M., Otto-Bliesner, B., Pagani, M., Pälike, H., Renssen, H., Royer, D.L., Siddall, M., Valdes, P., Zachos, J.C., and Zeebe, R.E.), Making sense of palaeoclimate sensitivity. *Nature, 491*, 683–691, 2012.

60. Martínez-Botí, M.A., Foster, G.L., Chalk, T.B., Rohling, E.J., Sexton, P.F., Lunt, D.J., Pancost, R.D., Badger, M.P.S., and Schmidt, D.N., Plio-Pleistocene climate sensitivity evaluated using high-resolution CO_2 records. *Nature, 518*, 49–53, 2015.

61. Hublin, J.J., Ben-Ncer, A., Bailey, S., Freidline, S.E., Neubauer, S., Skinner, M.M., Bergmann, I., La Cabec, A., Benazzi, S., Harvati, K., and Gunz, P., New fossils from Jebel Irhoud, Morocco and the pan-African origin of Homo sapiens. *Nature, 546*, 289–292, 2017.

Chapter 4

62. Royer, D.L., Berner, R.A., Montañez, I.P., Tabor, N.J., and Beerling, D.J., CO_2 as primary driver of Phanerozoic climate. *GSA Today, 14*, 4–10, 2004.

63. Cui, Y., Kump, L.R., Ridgwell, A.J., Charles, A., Junium, C.K., Diefendorf, A.F., Freeman, K.H., Urban, N.M., and Harding, I.C., Slow release of fossil carbon during the Palaeocene-Eocene Thermal Maximum. *Nature Geoscience, 4*, 481–485, 2011.

64. Zeebe, R.E., Ridgwell, A., and Zachos, J.C., Anthropogenic carbon release rate unprecedented during the past 66 million years, *Nature Geoscience, 9*, 325–329, 2016.

65. http://co2now.org/Current-CO2/CO2-Now/global-carbon-emissions.html
66. Suchet, P.A., Probst, J.-L., and Ludwig, W., Worldwide distribution of continental rock lithology: Implications for the atmospheric/soil CO_2 uptake by continental weathering and alkalinity river transport to the oceans, *Global Biogeochemical Cycles, 17*, 1038, doi:10.1029/2002GB001891, 2003.
67. http://www.pmel.noaa.gov/co2/story/What+is+Ocean+Acidification%3F
68. Caldeira, K., and Wickett, M.E., Oceanography: anthropogenic carbon and ocean pH. *Nature, 425*, p. 365, 2003.
69. Van Nes, E, Scheffer, M., Brovkin, V., Lenton, T.M., Ye, H., Deyle, E., and Sugihara, G., Causal feedbacks in climate change. *Nature Climate Change, 5*, 445–448, 2015.
70. Zeebe, R.E., Zachos, J.C., Caldeira, K., and Tyrrell, T., Carbon emissions and acidification. *Science, 321*, 51–52, 2008.
71. http://www.ipcc.ch/report/ar5/wg1/
72. Hansen, J., Sato, M., Kharecha, P., von Schukmann, K., Beerling, D.J., Cao, J., Marcott, S., Masson-Delmotte, V., Prather, M.J., Rohling, E.J., Shakun, J., Smith, P., Lacis, A., Russell, G., and Ruedy, R., Young people's burden: requirement of negative CO_2 emissions. *Earth System Dynamics, 8*, 577–616, 2017.
73. Sundquist, E.T., The global carbon dioxide budget. *Science, 259*, 934–941, 1993.
74. Ciais, P., Tagliabue, A., Cuntz, M., Bopp, L., Scholze, M., Hoffmann, G. Lourantou, A., Harrison, S.P., Prentice, I.C., Kelley, D.I., Koven, C., and Piao, S.L., Large inert carbon pool in the terrestrial biosphere during the Last Glacial Maximum. *Nature Geoscience, 5*, 74–79, 2012.
75. Martínez-Botí, M.A., Marino, G., Foster, G.L., Ziveri, P., Henehan, M.J., Rae, J.W.B., Mortyn, P.G., and Vance, D., Boron isotope evidence for oceanic carbon dioxide leakage during the last deglaciation. *Nature, 518*, 219–222, 2015.
76. Rohling, E.J., Marino, G., Foster, G.L., Goodwin, P.A., von der Heydt, A.S., and Köhler, P., Comparing climate sensitivity, past and present. *Annual Review of Marine Science, 10*, 261–288, 2018. http://www.annualreviews.org/doi/10.1146/annurev-marine-121916-063242
77. Imbrie, J., and Imbrie, K.P., *Ice ages: solving the mystery.* Harvard University Press, Cambridge, Massachusetts, 224 pp., 1986.
78. Milankovitch, M., Kanon der Erdbestrahlung und seine Anwendung auf das Eiszeitenproblem, *Special Publication 133, Mathematics and Natural Sciences Section.* Royal Serbian Academy, Belgrade, 1941.
79. Burton, M., and Salerno, G., Global volcanic CO_2 fluxes have been underestimated due to neglect of light scattering processes. *Geophysical Research Abstracts, 16*, EGU2014–11340, 2014.
80. Andres, R.J., and Kasgnoc, A.D., A time-averaged inventory of volcanic sulphur emissions. *Journal of Geophysical Research, 103*, 22251–22261, 1998.
81. http://earthobservatory.nasa.gov/Features/Volcano/

82. Luterbacher, J., and Pfister, C., The year without a summer. *Nature Geoscience,* 8, 246–248, 2015.
83. http://en.wikipedia.org/wiki/Supervolcano
84. http://en.wikipedia.org/wiki/Large_igneous_province
85. Vellekoop, J., Sluijs, A., Smit, J., Schouten, S., Weijers, J.W.H., Sinninghe Damsté, J.S., and Brinkhuis, H., Rapid short-term cooling following the Chicxulub impact at the Cretaceous-Paleogene boundary. *Proceedings of the National Academy of Sciences of the USA, 111,* 7537–7541, 2014.
86. Viera, L.E.A., Solanski, S.K., Krivova, N.A., and Usoskin, I. Evolution of the solar irradiance during the Holocene. *Astronomy and Astrophysics, 531,* A6, 2011. DOI: 10.1051/0004-6361/201015843.
87. Solanski, S.K., Krivova, N.A., and Haigh, J.D., Solar irradiance variability and climate. *Annual Review of Astronomy and Astrophysics, 51,* 311–351, 2013.
88. Maasch, K.A., Mayewski, P.A., Rohling, E.J., Stager, J.C., Karlen, W., Meeker, L.D., and Meyerson, E.A., A 2000 year context for modern climate change. *Geografiska Annaler, 87A,* 7–15, 2005.
89. Cini-Castagnoli, G., Bernasconi, S.M., Bonino, G., Della Monica, P., and Taricco, C., 700 Year record of the 11 year solar cycle by planktonic foraminifera of a shallow water Mediterranean core. *Advances in Space Research, 24,* 233–236, 1999.
90. https://www.imperial.ac.uk/media/imperial-college/grantham-institute/public/publications/briefing-papers/Solar-Influences-on-Climate---Grantham-BP-5.pdf

Chapter 5

91. http://www.oco.noaa.gov/roleofOcean.html
92. http://www.nodc.noaa.gov/OC5/3M_HEAT_CONTENT/
93. http://en.wikipedia.org/wiki/Ocean
94. http://wps.prenhall.com/wps/media/objects/7002/7170523/FG03_04.JPG
95. Roemmich, D., Church, J., Gilson, J., Monselesan, D., Sutton, P., and Wijffels, S., Unabated planetary warming and its ocean structure since 2006. *Nature Climate Change, 5,* 240–245, 2015.
96. Lyman, J.M., Good, S.A., Gouretski, V.V., Ishii, M., Johnson, G.C., Palmer, M.D., Smith, D.M., and Willis, J.K., Robust warming of the global upper ocean. *Nature, 465,* 334–337, 2010.
97. Resplandy, L., Keeling, R.F., Eddebbar, Y., Brooks, M.K., Wang, R., Bopp, L., Long, M.C., Dunne, J.P., Koeve W., and Oschlies, A., Quantification of ocean heat uptake from changes in atmospheric O_2 and CO_2 composition. *Nature,* 563, 105–108, 2018.

98. Figure SPM-05 in http://www.ipcc.ch/report/graphics/index.php?t= Assessment%20Reports&r=AR5%20-%20WG1&f=SPM

99. Goodwin, P.A., Katavouta, A., Roussenov, V.M., Foster, G.L., Rohling, E.J., and Williams, R.G., Pathways to 1.5°C and 2°C warming based on observational and geological constraints. *Nature Geoscience, 11*, 102–107, 2018.

100. Bousquet, P., Ciais, P., Miller, J.B., Dlugokencky, E.J., Hauglustaine, D.A., Prigent, C., Van der Werf, G.R., Peylin, P., Brunke, E.G., Carouge, C., Langenfelds, R.L., Lathière, J., Ramonet, M., Schmidt, M., Steele, L.P., Tyler, S.C., and White, J., Contribution of anthropogenic and natural sources to atmospheric methane variability. *Nature, 443*, 439–443, 2006.

101. http://www.grida.no/publications/other/ipcc_tar/?src=/climate/ipcc_tar/wg1/134.htm

102. http://www.esrl.noaa.gov/gmd/ccgg/trends/

103. http://www.esrl.noaa.gov/gmd/ccgg/trends/history.html

104. http://www.ipcc.ch/publications_and_data/ar4/wg1/en/tssts-2-1-1.html

105. Foster, G.L., and Rohling, E.J., Relationship between sea level and climate forcing by CO_2 on geological timescales. *Proceedings of the National Academy of Sciences of the USA, 110*, 1209–1214, 2013.

106. Chalk, T.B., Hain, M.P., Foster, G.L., Rohling, E.J., Sexton, P.F., Badger, M.P.S., Cherry, S.G., Hasenfratz, A.P., Haug, G.H., Jaccard, S.L., Martínez-García, A., Pälike, H., Pancost, R.D., and Wilson, P.A., Causes of ice-age intensification across the Mid-Pleistocene Transition. *Proceedings of the National Academy of Sciences of the USA, 114*, 13114–13119, 2017.

107. Huang, J., and McElroy, M.B., The contemporary and historical budget of atmospheric CO_2. *Canadian Journal of Physics, 90*, 707–716, 2012.

108. Stuiver, M., Burk, R.L., and Quay, P.D., $^{13}C/^{12}C$ ratios in tree rings and the transfer of biospheric carbon to the atmosphere. *Journal of Geophysical Research, 89*, 11731–11748, 1984.

109. Francey, R.J., Allison, C.E., Etheridge, D.M., Trudinger, C.M., Enting, I.G., Leuenberger, M., Langenfelds, R.L., Michel, E., and Steele, L.P., A 1000-year high precision record of $\delta^{13}C$ in atmospheric CO_2. *Tellus, 51B*, 170–193, 1999.

110. Quay, P.D., Tilbrook, B., and Wong C.S., Oceanic uptake of fossil fuel CO_2: carbon-13 evidence. *Science, 256*, 74–79, 1992.

111. Wei, G., McCulloch, M.T., Mortimer, G., Deng, W., and Xie, L., Evidence for ocean acidification in the Great Barrier Reef of Australia. *Geochimica et Cosmochimica Acta, 73*, 2332–2346, 2009.

112. Tans, P.P., de Jong, A.F.M., and Mook, W.G., Natural atmospheric ^{14}C variation and the Suess effect. *Nature, 280*, 826–828, 1979.

113. http://www.esrl.noaa.gov/gmd/outreach/isotopes/c14tellsus.html

114. Held, I.M., and Soden, B.J., Robust responses of the hydrological cycle to global warming. *Journal of Climate, 19*, 5686–5699, 2006.

115. Seidel, D.J., Fu, Q., Randel, W.J., and Reichler, T.J., Widening of the tropical belt in a changing climate. *Nature Geoscience, 1*, 21–24, 2008.
116. Isaac, J., and Turton, S., Expansion of the tropics: evidence and implications, Essay 5 in *State of the Tropics, 2014 report*, http://stateofthetropics.org/state-of-the-tropics-the-essays, 2014.
117. Parmesan, C., Ecological and evolutionary responses to recent climate change. *Annual Review of Ecology, Evolution, and Systematics, 37*, 637–669, 2006.
118. Poloczanska, E.S., Brown, C.J., Sydeman, W.J., Kiessling, W., Schoeman, D.S., Moore, P.J., Brander, K., Bruno, J.F., Buckley, L.B., Burrows, M.T., Duarte, C.M., Halpern, B.S., Holding, J., Kappel., C.V., O'Connor, M.I., Pandolfi, J.M., Parmesan, C., Schwing, F., Thompson, S.A., and Richardson, A.J., Global imprint of climate change on marine life. *Nature Climate Change, 3*, 919–925, 2013.
119. Hoegh-Guldberg, O., and Bruno, J.F. The impact of climate change on the world's marine ecosystems. *Science, 328*, 1523–1528, 2010.
120. Seimon, T.A., Seimon, A., Daszak, P., Halloy, S.R.P., Schloegel, L.M., Aguilar, C.A., Sowell, P., Hyatt, A.D., Konecky, B., and Simmons, J.E., Upward range extension of Andean anurans and chytridiomycosis to extreme elevations in response to tropical deglaciation. *Global Change Biology, 13*, 288–299, 2007.
121. De Vos, J.M., Joppa, L.N., Gittleman, J.L., Stephens, P.R., and Pimm, S.L., Estimating the normal background rate of species extinction. *Conservation Biology, 29*, 452–462, 2014.
122. Mora, C., Dousset, B., Caldwell, I.R., Powell, F.E., Geronimo, R.C., Bielecki, C.R., Counsell, C.W.W., Dietrich, B.S., Johnston, E.T., Louis, L.V., Lucas, M.P., McKenzie, M.M., Shea, A.G., Tseng, H., Giambelluca, T.W., Leon, L.R., Hawkins, E., and Trauernicht, C., Global risk of deadly heat. *Nature Climate Change, 7*, 501–506, 2017.
123. https://cosmosmagazine.com/geoscience/perfect-storm-threatens-the-world-s-reefs
124. Hughes, T.P., Anderson, K.D., Connolly, S.R., Heron, S.F., Kerry, J.T., Lough, J.M., Baird, A.H., Baum, J.K., Berumen, M.L., Bridge, T.C., Claar, D.C., Eakin, C.M., Gilmour, J.P., Graham, N.A.J., Harrison, H., Hobbs, J.P.A., Hoey, A.S., Hoogenboom, M., Lowe, R.J., McCulloch, M.T., Pandolfi, J.M., Pratchett, M., Schoepf, V., Torda, G., and Wilson, S.K., Spatial and temporal patterns of mass bleaching of corals in the Anthropocene. *Science, 359*, 80–83, 2018.
125. Laliberté, F., Zika, J., Mudryk, L., Kushner, P.J., Kjellsson, J., and Döös, K., Constrained work output of the moist atmospheric heat engine in a warming climate. *Science, 347*, 540–543, 2015.
126. http://www.aoml.noaa.gov/hrd/tcfaq/tcfaqHED.html
127. https://oceantoday.noaa.gov/hurricanestormsurge/
128. Kossin, J.P., Emanuel, K.A., and Vecchi, G.A., The poleward migration of the location of tropical cyclone maximum intensity. *Nature, 509*, 349–352, 2014.

129. Lucas, C., Timbal, B., and Nguyen, H., The expanding tropics: a critical assessment of the observational and modeling studies. *WIREs Climate Change*, 5, 89–112, 2014.

130. Haarsma, R.J., Hazeleger, W., Severijns, C., de Vries, H., Sterl, A., Bintanja, R., van Oldenburgh, G.J., and van den Brink, H., More hurricanes to hit western Europe due to global warming. *Geophysical Research Letters*, 40, 1783–1788, 2013.

131. Emanuel, K., Increasing destructiveness of tropical cyclones over the past 30 years. *Nature*, 436, 686–688, 2005.

132. ftp://texmex.mit.edu/pub/emanuel/PAPERS/Haurwitz_2008.pdf

133. Westbrook, G.K., Thatcher, K.E., Rohling, E.J., Piotrowski, A.M., Pälike, H., Osborne, A.H., Nisbet, E.G., Minshull, T.A., Lanoisellé, M., James, R.H., Hühnerbach, V., Green, D., Fisher, R.E., Crocker, A.J., Chabert, A., Bolton, C., Beszczynska-Möller, A., Berndt, C., and Aquilina, A., Escape of methane gas from the seabed along the West Spitsbergen continental margin. *Geophysical Research Letters*, 36, L15608, doi:10.1029/2009GL 039191, 2009.

134. http://www.swerus-c3.geo.su.se/index.php/swerus-c3-in-the-media/news/177-swerus-c3-first-observations-of-methane-release-from-arctic-ocean-hydrates

135. http://www.realclimate.org/index.php/archives/2014/08/how-much-methane-came-out-of-that-hole-in-siberia/

136. Mann, M.E., Rahmstorf, S., Kornhuber, K., Steinman, B.A., Miller, S.K., and Coumou, D., Influence of anthropogenic climate change on planetary wave resonance and extreme weather events. *Scientific Reports*, 7, 45242, 2017. doi:10.1038/srep45242

137. Kretschmer, M., Coumou, D., Agel, L., Barlow, M., Tziperman, E., and Cohen, J., More-persistent weak stratospheric polar vortex states linked to cold extremes. *Bulletin of the American Meteorological Society, January 2018*, 49–60, 2018. doi: 10.1175/BAMS-D-16-0259.1

138. Forzeri, G., Cescatti, A., Batista e Silva, F., and Feyen, L., Increasing risk over time of weather-related hazards to the European population: a data-driven prognostic study. *Lancet Planet Health*, 1, e200–e208, 2017.

139. Hoegh-Guldberg, O., Mumby, P.J., Hooten, A.J., Steneck, R.S., Greenfield, P., Gomez, E., Harvell, C.D., Sale, P.F., Edwards, A.J., Caldeira, K., Knowlton, N., Eakin, C.M., Iglesias-Prieto, R., Muthiga, N., Bradbury, R.H., Dubi, A., and Hatziolos, M.E., Coral reefs under rapid climate change and ocean acidification. *Science*, 318, 1737–1742, 2007.

140. Veron, J.E., Hoegh-Guldberg, O., Lenton, T.M., Lough, J.M., Obura, D.O., Pearce-Kelly, P., Sheppard, C.R.C., Spalding, M., Stafford-Smith M.G., and Rogers, A.D., The coral reef crisis: The critical importance of <350 ppm CO_2. *Marine Pollution Bulletin*, 58, 1428–1436, 2009.

141. Kroeker, K.J., Kordas, R.L., Crim, R., Hendriks, I.E., Ramajo, L., Singh, G.S., Duarte, C.M., and Gattuso, J.P., Impacts of ocean acidification on marine organisms: quantifying sensitivities and interaction with warming. *Global Change Biology, 19*, 1884–1896, 2013.

142. Ekstrom, J.A., Suatoni, L., Cooley, S., Pendleton, L.H., Waldbusser, G.G., Cinner, J.E., Ritter, J., Langdon, C., van Hooidonk, R., Gledhill, D., Wellman, K., Beck, M.W., Brander, L.M., Rittschof, D., Doherty, C., Edwards, P.E.T., and Pertela, R., Vulnerability and adaptation of US shellfisheries to ocean acidification. *Nature Climate Change, 5*, 207–214, 2015.

143. O'Dea, S.A., Gibbs, S.J., Brown, P.R., Young, J.R., Poulton, A.J., Newsam, C., and Wilson, P.A., Coccolithophore calcification response to past ocean acidification and climate change. *Nature Communications, 5*, 7 pp., DOI: 10.1038/ncomms6363, 2014.

144. Dessler, A., & Forster, P., An estimate of equilibrium climate sensitivity from interannual variability, February 6, 2018. Retrieved from eartharxiv.org/4et67

145. Cox, P.M., Huntingford, C., and Williamson, M.S., Emergent constraint on equilibrium climate sensitivity from global temperature variability. *Nature* 553, 319–322, 2018.

146. Copenhagen Accord (2009) United Nations Framework Convention on Climate Change, Draft decision 2/CP.15 FCCC/CP/2009/L.7 18 December 2009.

147. Randalls, S., History of the 2°C climate target. *WIREs Climate Change, 1*, 598–605, 2010.

148. International Symposium on the stabilisation of greenhouse gas concentrations, Report of the International Scientific Steering Committee. Met Office. 10 May 2005. Retrieved 15 March 2007. http://www.g8.utoronto.ca/environment/2005steeringcommittee.pdf

149. Marcott, S.A., Shakun, J.D., Clark, P.U., and Mix, A.C., A reconstruction of regional and global temperature for the past 11,300 years. *Science, 339*, 1198–1201, 2013.

150. World Bank, *Turn down the heat: climate extremes, regional impacts, and the case for resilience—full report.* Washington D.C., 2013. http://documents.worldbank.org/curated/en/2013/06/17862361/turn-down-heat-climate-extremes-regional-impacts-case-resilience-full-report

151. McGlade, C., and Ekins, P., The geographical distribution of fossil fuels unused when limiting global warming to 2°C. *Nature, 517*, 187–190, 2015.

152. Pritchard, H.D., Ligtenberg, S.R.M., Fricker, H.A., Vaughan, D.G., van den Broeke, M.R., and Padman, L., Antarctic ice-sheet loss driven by basal melting of ice shelves. *Nature, 484*, 502–505, 2012.

153. Khan, S.A., Kjær, K.H., Bevis, M., Bamber, J.L., Wahr, J., Kjeldsen, K.K., Bjørk, A.A., Korsgaard, N.J., Stearns, L.A., van den Broeke, M.R., Liu,

L., Larsen, N.K., and Muresan, I.S., Sustained mass loss of the northeast Greenland ice sheet triggered by regional warming. *Nature Climate Change,* 4, 292–299, 2014.
154. Favier, L., Durand, G., Cornford, S.L., Gudmundsson, G.H., Gagliardini, O., Gillet-Chaulet, F., Zwinger, T., Payne, A.J., and Le Brocq, A.M., Retreat of Pine Island Glacier controlled by marine ice-sheet instability. *Nature Climate Change,* 4, 117–121, 2014.
155. Pollard, D., DeConto, R.M., and Alley, R.B., Potential Antarctic ice sheet retreat driven by hydrofracturing and ice cliff failure. *Earth and Planetary Science Letters, 412,* 112–121, 2015.
156. DeConto, R.M., and Pollard, D., Contribution of Antarctica to past and future sea-level rise. *Nature, 531,* 591–597, 2016.
157. Bindoff, N., Rintoul, S., and Haward, M., Position analysis: climate change and the Southern Ocean. *ACE CRC Oceans Position Analysis, 2011,* ISBN 978-0-9871939-0-2, 2011.
158. http://www.antarcticglaciers.org/glaciers-and-climate/ice-ocean-interactions/marine-ice-sheets/
159. Bougamont, M., Christoffersen, P., Hubbard, A.L., Fitzpatrick, A.A., Doyle, S.H., and Carter, S.P., Sensitive response of the Greenland Ice Sheet to surface melt drainage over a soft bed. *Nature Communications, 5,* 5052, doi:10.1038/ncomms6052, 2014.
160. Mengel, M., and Levermann, A., Ice plug prevents irreversible discharge from East Antarctica. *Nature Climate Change,* 4, 451–455, 2014.
161. Greenbaum, J.S., Blankenship, D.D., Young, D.A., Richter, T.G., Roberts, J L., Aitken, A.R.A., Legresy, B., Schroeder, D.M., Warner, R.C., van Ommen, T.D., and Siegert, M.J., Ocean access to a cavity beneath Totten Glacier in East Antarctica. *Nature Geoscience, 8,* 294–298, 2015.
162. Rohling, E.J, Haigh, I.D., Foster, G.L., Roberts, A.P., and Grant, K.M., A geological perspective on potential future sea-level rise. *Scientific Reports, 3,* 3461, doi:10.1038/srep03461, 2013.
163. Church, J.A., and White, N.J., Sea-level rise from the late 19th to the early 21st Century. *Surveys in Geophysics, 32,* 585–602, 2011.
164. Jevrejeva, S., Moore, J.C., Grinsted, A., and Woodworth, P.L., Recent global sea level acceleration started over 200 years ago? *Geophysical Research Letters, 35,* L08715, doi:10.1029/2008GL033611, 2008.
165. Rignot, E., Velicogna, I., van den Broeke, M.R., Monaghan, A., and Lenaerts, J., Acceleration of the contribution of the Greenland and Antarctic ice sheets to sea level rise. *Geophysical Research Letters, 3,* L05503, doi:10.1029/2011GL046583, 2011.
166. Goodwin, P., Haigh, I.D., Rohling, E.J., and Slangen, A., A new approach to projecting 21st century sea-level changes and extremes. *Earth's Future, 5,* 240–253, 2017.

167. Hanson, S., Nicholls, R., Ranger, N., Hallegatte, S., Corfee-Morlot, J., Herweijer, C., and Chateau, J., A global ranking of port cities with high exposure to climate extremes. *Climatic Change, 104,* 89–111, 2011.

Chapter 6

168. http://www.skepticalscience.com/weathering.html
169. http://en.wikipedia.org/wiki/Enhanced_weathering
170. Liu, Y.Y., van Dijk, A.I.J.M., de Jeu, R.A.M., Canadell, J.G., McCabe, M.F., Evans, J.P., and Wang, G., Recent reversal in loss of global terrestrial biomass. *Nature Climate Change, 5,* 470–474, 2015.
171. Erb, K.H., Kastner, T., Plutzar, C., Bais, A.L., Carvalhais, N., Fetzel, T., Ginrich, S., Haberl, H., Lauk, C., Niedertscheider, M., Pongratz, J., Thurner, M., and Luyssaert, S., Unexpectedly large impact of forest management and grazing on global vegetation biomass. *Nature, 553,* 73–76, 2018.
172. Larrasoaña, J.C., Roberts, A.P., and Rohling, E.J., Dynamics of green Sahara periods and their role in hominin evolution. *PLoS ONE, 8,* e76514, doi:10.1371/journal.pone.0076514, 2013.
173. http://www.highstand.org/erohling/DarkMed/dark-title.html
174. Rohling, E.J., Marino, G., and Grant, K.M., Mediterranean climate and oceanography, and the periodic development of anoxic events (sapropels). *Earth Science Reviews, 143,* 62–97, 2015.
175. De Lange, G.J., Thomson, J., Reitz, A., Slomp, C.P., Speranza Principato, M., Erba, E., and Corselli, C., Synchronous basin-wide formation and redox-controlled preservation of a Mediterranean sapropel. *Nature Geoscience, 1,* 606–610, 2008.
176. Meijer, P.Th., and Krijgsman, W., A quantitative analysis of the dessication and re-filling of the Mediterranean during the Mediterranean Salinity Crisis. *Earth and Planetary Science Letters, 240,* 510–520, 2005.
177. http://wps.prenhall.com/wps/media/objects/7002/7170523/FG03_04.JPG
178. Peters, G.P., and Geden, O., Catalysing a political shift from low to negative carbon. *Nature Climate Change, 7,* 619–621, 2017.
179. Cressey, D., Geochemistry: Rock's power to mop up carbon revisited. *Nature, 505,* p. 464, 2014.
180. http://en.wikipedia.org/wiki/Dunite
181. http://topochange.cr.usgs.gov/ranking.php
182. http://www.world-nuclear.org/information-library/energy-and-the-environment/clean-coal-technologies.aspx
183. http://avoid-net-uk.cc.ic.ac.uk/wp-content/uploads/delightful-downloads/2015/07/Planetary-limits-to-BECCS-negative-emissions-AVOID-2_WPD2a_v1.1.pdf

184. Kato, E., and Yamagata, Y., BECCS capability of dedicated bioenergy crops under a future land-use scenario targeting net negative carbon emissions, *Earth's Future, 2*, 421–439, 2014.

185. Williamson, P., Scrutinize CO_2 removal methods. *Nature, 530*, 153–155, 2016.

186. Griscom, B.W., Adams, J., Ellis, P.W., Houghton, R.A., Lomax, G., Miteva, D.A., Schlesinger, W.H., Shoch, D., Siiämaki, J.V., Smith, P., Woodbury, P., Zganjar, C., Blackman, A., Campari, J., Conant, R.T., Delgado, C., Elias, P., Gopalakrishna, T., Hamsik, M.R., Herrero, M., Kiesecker, J., Landis, E., Laestadius, L., Leavitt, S.M., Minnemeyer, S., Polasky, S., Potapov, P., Putz, F.E., Sanderman, J., Silvius, M., Wollenberg, E., and Fargione, J., Natural climate solutions. *Proceedings of the National Academy of Sciences of the USA, 114*, 11645–11650, 2017.

187. Beerling, D.J., Leake, J.R., Long, S.P., Scholes, J.D., Ton, J., Nelson, P.N., Bird, M., Kantzas, E., Taylor, L.L., Sarkar, B., Kelland, M., DeLucia, E., Kantola, I., Muller, C., Rau, G., and Hansen, J., Farming with crops and rocks to address global climate, food and soil security. *Nature Plants, 4*, 138–147, 2018.

188. Renforth, P., and Henderson, G., Assessing ocean alkalinity for carbon sequestration. *Reviews of Geophysics, 55*, 636–674, 2017.

189. Ten Berge, H.F.M., van der Meer, H.G., Steenhuizen, J.W., Goedhart, P.W., Knops, P., and Verhagen, J., Olivine weathering in soil, and its effects on growth and nutrient uptake in ryegrass (*Lolium perenne* L.): a pot experiment. *PLOS ONE, 7*, e42098, 2012.

190. Fasihnikoutalab, M.H., Westgate, P., Huat, B.K.K., Asadi, A., Ball, R.J., Nahazanan, H., and Singh, P., New insights into potential capacity of olivine in ground improvement. *Electronic Journal of Geotechnical Engineering, 20*, 2137–2148, 2015.

191. http://www.drawdown.org

Chapter 8

192. Kelley, C.P., Mohtadi, S., Cane, M.A., Seager, R., and Kushnir, Y., Climate change in the Fertile Crescent and implications of the recent Syrian drought. *Proceedings of the National Academy of Sciences of the USA, 112*, 3241–3246, 2015.

193. Weninger, B., Clare, L., Rohling, E.J., Bar-Yosef, O., Boehner, U., Budja, M., Bundschuh, M., Feurdean, A., Gebel, H.G., Joeris, O., Lindstaedter, J., Mayewski, P., Muehlenbruch, T., Reingruber, A., Rollefson, G., Schyle, D., Thissen, L., Todorova, H., and Zielhofer, C., The impact of rapid climate change on prehistoric societies during the Holocene in the Eastern Mediterranean. *Documenta Praehistorica, 36* (Neolithic Studies 16), 7–59, 2009.

194. Medina-Elizalde, M., and Rohling, E.J., Collapse of classic Maya civilization related to modest reduction in precipitation. *Science*, 335, 956–959, 2012.
195. Cline, E.H., *1177 B.C.: The year civilization collapsed*. Princeton University Press, Princeton, U.S.A., ISBN978-0691-14089-6, 237 pp., 2014.

GLOSSARY

albedo   measure of reflectivity

anthropogenic   human-caused (or human-generated)

astronomical cycle of eccentricity   variations in the shape of Earth's orbit around the Sun, from near circular to elliptical

astronomical cycle of obliquity   the gradually changing angle—or tilt—of Earth's rotational axis relative to the perpendicular to the plane of Earth's orbit

astronomical cycle of precession   the "wobble"—like a spinning top—of Earth's rotational axis relative to the plane of its orbit around the Sun, which causes the equinoxes and solstices to shift along the orbit

billion   one thousand million

biomass   biological material from living or recently living organisms

biosphere   the combination of all life

carbon cycle   the complex interactions that control carbon storage and exchange between the biosphere (life), hydrosphere (oceans, lakes, rivers), and lithosphere (rocks and sediments)

carbonate compensation   the interaction between carbonate in deep-sea sediment and the deep-water chemistry, whereby an increase in deep-water CO_2 is "buffered" (negative feedback) by dissolution of sedimentary carbonate, and a decrease in deep-water CO_2 is buffered by increased preservation of sedimentary carbonate

climate sensitivity   temperature change in degrees Centigrade (°C) per W/m^2 of radiative forcing of climate

core-top calibration   practice of comparing proxy data measured in recent sediments (sediment core tops) with the represented properties of sea water above those core tops

cosmogenic radionuclides   specific elemental isotopes that are formed by inter-action of galactic rays with Earth's upper atmosphere

Cretaceous Period   interval of geological time, about 66 to 144 million years ago

deglaciation   termination of an ice age, with large-scale decay of ice sheets

Earth's energy balance   the balance between how much energy comes in versus how much goes out, as measured at the top of the atmosphere

Eocene Epoch   interval of geological time, about 34 to 56 million years ago

external carbon   carbon in a reservoir—typically sediments or fossil fuels—that is external to the hydrosphere-biosphere-atmosphere system

feedback   a response that amplifies (positive feedback) or dampens (negative feedback) the impact of an initial change or disturbance

foraminifera   single-celled organisms with a calcium carbonate shell, which are very abundant in the ocean, both free-floating in surface waters (planktonic) and resident on the sea floor (benthic)

gas-hydrate   crystalline water-based solids that look like ice, which trap gases inside "cages" of hydrogen-bonded water molecules

giga   one billion (one thousand million)

greenhouse gas   gas in the atmosphere that absorbs and emits radiation within the thermal infrared range

hydrosphere   the system of fluid water (oceans, lakes, rivers, groundwater). In the broadest sense, this would include the water present in the form of ice, but in climatology that is commonly separated out under the name cryosphere

industrial revolution   the transition to new processes in manufacturing, from about 1760 into the early 1800s, which included the widespread introduction of coal-generated power and continued thereafter on the basis of combined coal, oil, and gas

infrared   light of longer wavelengths (700 to about 1000 nanometers), which humans normally cannot see

kilo   one thousand

Last Glacial Maximum   maximum of the most recent major ice age, roughly 20 thousand years ago

lithosphere   in simplest terms, this is the total of rocks and sediments

Little Ice Age   a generally cool period, roughly spanning the years 1400 to 1850, but with notable climatic fluctuations within that period

Maunder Minimum   period of minimum sunspot activity between about 1645 and 1715

milli   one thousandth (one thousand milligrams in a gram)

million   one thousand thousand

moraine   accumulation of material that has either fallen onto a glacier surface, or has been pushed along by a glacier

nahcolite   sodium-carbonate mineral ($NaHCO_3$)

nano   one billionth (one billion nanometers in a meter)

paleoceanography   research on past oceans
Paleocene Epoch   interval of geological time, about 56 to 66 million years ago
paleoclimate   past climate
paleoclimatology   research on past climates
plate tectonics   the slow movement of continental plates around the world
Pliocene Epoch   interval of geological time, about 2.6 to 5.3 million years ago
polar amplification   the ratio between polar temperature change and global average temperature change
ppb   parts per billion
ppm   parts per million
pre-industrial   before the industrial revolution
proxy   indirect measure of a variable climate or ocean property
radiocarbon   a radioactively decaying isotope of carbon (carbon-14)
salinity   salt-content, notably of sea water
sapropel   commonly dark-colored sediments that are rich in organic matter
saturation state   how much of a certain substance is contained—for example in ocean water—relative to the maximum amount that could be contained under the same conditions
sediment   material that is broken down by weathering and erosion and then transported by wind, water, or ice, or simply falling due to gravity
spectral absorption   absorption of light/radiation in specific spectral bands (specific wavelengths)
stomata   openings, comparable to pores, in leaves that are used to regulate a plant's gas and vapor exchange with the atmosphere
subduction zone   plate-tectonic term for a zone where a marine plate with sediments is pushed underneath a (commonly) continental plate
Triassic Period   interval of geological time, 201 to 252 million years ago
trillion   one million million
ultraviolet   light of short wavelengths (10 to 400 nanometers) that humans cannot see
watt   unit of power (equivalent to a joule per second)

ABBREVIATIONS

CH_4 methane
CIE   carbon isotope event
CO_2 carbon dioxide
CS   climate sensitivity
ECS   equilibrium climate sensitivity
ESS   Earth system sensitivity
GtC   gigaton of carbon

GtS   gigaton of sulfur
IPCC   Intergovernmental Panel for Climate Change
ISWR   incoming short-wave radiation
OLWR   outgoing long-wave radiation
PETM   Palaeocene-Eocene Thermal Maximum (about 56 million years ago)
pH   chemical measure of acidity or alkalinity (negative logarithm of the hydrogen ion activity in water)
SO_2 sulfur dioxide
TSI   Total Solar Irradiance

INDEX

Tables, figures, and boxes are indicated by an italic *t*, *f*, and *b*, respectively, following the page number.

atmosphere
on Earth's temperature, 42
in greenhouse effect, 42, 50–51
Australia, green energy
opportunities, 130–31

benthic foraminifera
boron/calcium ratios, 26
oxygen isotope ratios, sea-level change
data, 29, 30–32
bicycle, feedback, 46, 66–67
bioenergy with carbon capture and storage
(BECCS), 123
biosphere, terrestrial. *See* terrestrial
biosphere
boron/calcium ratios, benthic
foraminifera, 26
boron isotopes, microfossil, 14*f*, 25, 26
box-models, 11
burps, carbon (CO_2 and CH_4), 56, 57
carbon isotope events, 2*f*, 3–4*f*, 59–61
Paleocene-Eocene Thermal Maximum,
3–4*f*, 59–61, 119*b*
Paleocene-Eocene Thermal Maximum,
clean-up after, 119*b*

carbon, external
in climate change, 126
definition, 23–24
emissions (*see* carbon dioxide (CO_2)
emissions)
exchanges, 55
gigatons, 53–55, 54*f*
human input and impact, 3–4*f*, 54*f*, 56
hydrosphere-biosphere-atmosphere
system, 23–24, 28
input-output balance and fluctuations,
55–56, 63*f*
input-output net loss, minor, 54*f*, 55–56
inputs, CO_2 and CH_4 burps or human-
caused emissions, 54*f*, 56, 57, 63*f*
subduction margins, 56, 57
carbonate compensation, 26–28, 66
carbonate rocks. *See also* lithosphere
weathering, 57–58
carbon burial, CO_2 removal by,
61–62, 115–17

carbon capture, human intervention, 120
carbon capture and storage (CCS), 122–23
bioenergy with, 123
carbon cycle. *See also specific components*
carbon dioxide, 45, 53
gigatons, carbon, 53–55, 54*f*
greenhouse effect, 45
greenhouse effect, slow components
on, 46
methane, 45, 53
carbon cycle, changes, 53–67, 54*f*, 60*b*, 63*f*
balance failure, 58, 60*b*
before human impact, natural
circumstances, 55
burps, 54*f*, 56, 57, 63*f*
burps, carbon isotope events, 2*f*,
3–4*f*, 59–61
burps, Paleocene-Eocene Thermal
Maximum, 3–4*f*, 59–61, 119*b*
carbon exchange, hydrosphere-biosphere-
atmosphere system, 56–58, 60*b*
flux, 53–55, 54*f*
human impact, external carbon input,
3–4*f*, 54*f*, 56
input-output balance and fluctuations,
55–56, 63*f*
life, 58, 60*b*
lithosphere carbon exchange, 56–58
plate tectonics, 57–58
redistributions, 53, 61–67 (*see also*
redistributions, carbon cycle)
reservoirs, 54*f*, 55
reservoirs, carbon fluxes between, 54*f*,
55, 63*f*
reservoirs, volume, 53–55, 54*f*
timescales, 14*f*, 53, 54*f*, 55–56, 58,
61–62, 63*f*
timescales, ocean mixing, 62, 64
timescales, ocean saturation state, 64
timescales, processes on, 58, 63*f*
timescales, slow feedbacks, 66–67
timescales, terrestrial biosphere, 65–66
web of interactions, 53–55, 54*f*
carbon dioxide (CO_2), atmospheric
carbon cycle, 45, 53
as driver of change and carbon cycle
feedbacks, 45–46, 51